FROM IVF TO IMMORTALITY

FROM IVF TO IMMORTALITY

Controversy in the Era of Reproductive Technology

Ruth Deech

Anna Smajdor

OXFORD
UNIVERSITY PRESS

Great Clarendon Street, Oxford OX2 6DP

Oxford University Press is a department of the University of Oxford.
It furthers the University's objective of excellence in research, scholarship,
and education by publishing worldwide in

Oxford New York

Auckland Cape Town Dar es Salaam Hong Kong Karachi
Kuala Lumpur Madrid Melbourne Mexico City Nairobi
New Delhi Shanghai Taipei Toronto

With offices in

Argentina Austria Brazil Chile Czech Republic France Greece
Guatemala Hungary Italy Japan Poland Portugal Singapore
South Korea Switzerland Thailand Turkey Ukraine Vietnam

Published in the United States
by Oxford University Press Inc., New York

First published 2007

British Library Cataloguing in Publication Data
Data available

Library of Congress Cataloging in Publication Data
Deech, Ruth.
From IVF to immortality: controversy in the era of reproductive technology / Ruth Deech, Anna Smajdor.
p.; cm.
Includes bibliographical references and index.
ISBN-13: 978-0-19-921978-0 (hbk.: alk. paper)
ISBN-13: 978-0-19-921979-7 (pbk.: alk. paper) 1. Great Britain. Human Fertilisation & Embryology Authority. 2. Human reproductive technology—Moral and ethical aspects—Great Britain. 3. Human reproductive technology—Law and legislation—Great Britain. 4. Human reproductive technology—Social aspects—Great Britain. 5. Human reproduction—Government policy—Great Britain.
[DNLM: 1. Great Britain. Human Fertilisation & Embryology Authority. 2. Reproductive Techniques, Assisted—legislation & jurisprudence—Great Britain. 3. Embryo Research—ethics—Great Britain. 4. Embryo Research—legislation & jurisprudence—Great Britain. 5. Reproductive Techniques, Assisted—ethics—Great Britain. WQ 33 FA1 D311f 2007] I. Samjdor, Anna. II. Title.
RG133.5.D43 2007
176—dc22 2007032630

Typeset by Newgen Imaging Systems (P) Ltd., Chennai, India
Printed in Great Britain
on acid-free paper by
Biddles Ltd., King's Lynn

ISBN 978-0-19-921978-0 (Hbk.) 978-0-19-921979-7 (Pbk.)

10 9 8 7 6 5 4 3 2 1

Acknowledgements

For my family. *Ruth Deech*

Thanks to Richard and Fiona Forsyth and Paul Ireland for their helpful comments and suggestions during the writing of this book. *Anna Smajdor*

Acknowledgements

Contents

Abbreviations xi
List of Illustrations xiii

Introduction 1
I. The legal framework of the Human Fertilisation and Embryology Act 3
II. Resources, to fight and to enforce 3
III. The power of the media 4
IV. Politics 4
V. Ethics 4
The ongoing need for regulation 5

1. Reproductive Technologies and the Birth of the HFEA 7
Natural reproduction: what is 'supposed' to happen, and why does it go wrong? 11
The process 12
When problems arise 13
Artificial insemination (AI) 15
In vitro fertilization (IVF) 17
Surplus embryos 20
Risks of IVF 20
Intracytoplasmic Sperm Injection (ICSI) 21
Preimplantation genetic diagnosis (PGD) and preimplantation genetic screening (PGS) 22
Out of the freezer ... 24
Cloning 25
Therapeutic cloning and stem cell research 28

2. Ethics, Embryos, and Infertility 29
Relief of infertility as a worthy goal 30
Ethics, law, and regulation 32
Respect for autonomy 34
Autonomy and safety 35
Autonomy and the public good 38

Autonomy and consent 39
Autonomy and privacy 42
Parental autonomy and harm to children 43
Health risks to children 44
Family setup and the welfare of the child 47
The non-identity problem 49

3. Saviour Siblings, Designer Babies, and Sex Selection 53
Preimplantation Genetic Diagnosis (PGD) 53
Positive selection 58
Screening for carriers 61
Sex selection 62
Saviour siblings 65

4. Fertility is a Feminist Issue 77
Equal interests in reproduction? 78
Egg freezing 81
The older mother 81
Risks associated with fertility treatment 87
Women's autonomy and fertility treatment 91
Future developments 95
Artificial gametes 95
Artificial wombs 97

5. Private Lives and Public Policy—The Story of Diane Blood 101
Consent 103
Does it matter what happens when we are irretrievably unconscious or dead? 104
Informed consent 106
Dignity 108
Is there a doctrine of 'best interests'? 109
Clinical and other interests 110
The spouse's testimony 112
Some further implications 117
Moral responsibility and the media 123

6. Human Rights and Reproduction 127
Prisoners and the right to reproduce 127
The rights in question 130
The extreme view: the right to reproductive choice 134
The right to privacy and family life: does it entail a right to reproduce? 135

Positive and negative rights 138
Rights and duties 138
The right to marry 140
Rights and regulation 141
The individual and society 142
A perverse incentive? 148
The downside of rights and European law 149
Rights and the precautionary principle 150
Homogeneity and national integrity 152

7. Deconstructing The Family 153
Donated gametes 155
Different views on anonymity 157
Babies with two mothers? 158
Surrogacy 159
Single parents: no need for a father ... or a mother? 165
Non-discrimination 172
The role of empirical data 174
Same sex parents 175
Changing values 177

8. Embryonic Stem Cells and Therapeutic Cloning 179
What are embryonic stem cells? 182
Foetal stem cells 186
Adult stem cells 187
Ethical concerns 188
Loss of potential 190
'Personhood' and special attributes 192
Social harms 194
Obtaining eggs 196
Do the benefits outweigh the harm? 198
Global concerns 199
Broadening the scope 200
Legal challenges 201
Other ways of obtaining stem cells 203
Single cell biopsy 203
Dead embryos 204
Parthenogenesis 204
The role of the HFEA 206
National differences in legislation/regulation 206
A. Germany 206

B. Italy 209
C. The USA 209
How do the various approaches compare? 211

Afterword 215

Notes 217
Index 231

Abbreviations

AI	artificial insemination
ART	assisted reproductive technology
ESC	embryonic stem cell
GM	genetically modified
HFE Act	Human Fertilisation and Embryology Act 1990
HFEA	Human Fertilisation and Embryology Authority
HLA	human leukocyte antigen
ICSI	intracytoplasmic sperm injection
IVF	*in vitro* fertilization
NHS	National Health Service
OHSS	ovarian hyperstimulation syndrome
PGD	preimplantation genetic diagnosis
PGS	preimplantation genetic screening
VLA	Voluntary Licensing Authority

List of Illustrations

1. A collection of early human embryos of different stages. © M Johnson, Wellcome Images. 13
2. Human eight cell embryo for IVF selection–Grade one. © K. Hardy, Wellcome Images. 18
3. Human eight cell embryo for IVF selection–Grade four. © K. Hardy, Wellcome Images. 19
4. A human embryo at day 3. © Yorgos Nikas, Wellcome Images. 23
5. Dolly cloning. © Wellcome Photo Library. 27
6. The blastocyst is a hollow ball. © Alan Handyside, Wellcome Images. 183
7. Therapeutic cloning. © Wellcome Photo Library. 185

Introduction

This book is an account of the radical developments in IVF (*in vitro* fertilization) and embryology over the last two decades. It explores the impact of these developments on concepts of parenthood and the family, and the ways in which these social and scientific issues are dealt with in the UK's legal and regulatory framework.

The Human Fertilisation and Embryology Act 1990 governs the use of sperm, eggs and embryos in research and fertility treatments. Under that law the power to regulate is given to the Human Fertilisation and Embryology Authority (HFEA). It is often asked by what moral right the members of the HFEA pronounce on these issues. The answer is that it embodies the democratic compromise between strongly held views in society, a compromise settled by provisions of the Human Fertilisation and Embryology Act 1990. The HFEA works within the Act to reconcile opposing views and point to a way forward, with public accountability.

The HFEA has many tasks. It licenses and monitors clinics that carry out IVF treatments, donor insemination and embryo research. It ensures that treatment and research are undertaken with respect for human life and responsibility towards the parties involved, recognizing the vulnerability of patients and the expensive nature of the treatments. One in seven couples seek infertility treatment and there may be many more who are infertile but do not seek medical advice. The statute and the Code of Practice drafted under the HFEA's powers embody the notion that care should be taken not to exploit these patients.

The HFEA regulates the storage of gametes, registers information about donors, treatments and children, and safeguards the biggest database of its kind in the world. It issues a Code of Practice to clinics, gives advice and information to patients, donors, clinics and the government, and keeps new developments under review. It has staff and premises in London, and twenty members selected openly on merit after advertisement. It issues an annual report, holds open meetings and keeps in touch with all elements of the relevant professions. Its overall aim is to ensure that public understanding and reassurance move at the same pace as the new developments that it licenses.

Recent decisions concerning fertility treatment under the HFE Act have been affected by the introduction into UK law of the European Human Rights Convention by the Human Rights Act 1998. This gives greater weight to individual wishes. Nevertheless human rights law does not support an absolute right to have children nor to be supplied with medical assistance to do so. The HFEA has faced several legal challenges, all but one of which have failed.[1]

In the sixteen years of its existence, the HFEA has overseen major steps forward in infertility treatment, extending from 'simple' IVF into matters of convenience, lifesaving and life alteration. Fears that allowing IVF or embryo research would inevitably lead down a slippery slope have largely been allayed. As far as is known, no one has tried to keep an embryo *in vitro* for more than fourteen days from fertilization, or attempted reproductive cloning, because policing of laboratories is part of the system.

Legal regulation in Britain has probably served to protect clinicians and scientists not only from legal action for malpractice (where they have followed the HFEA Code of Practice) but also to give them a shield against accusations of ethical malpractice, as long as they act within the parameters agreed by Parliament and the HFEA. These safeguards clearly do not apply where there has been a total breakdown in the clinic controls and deliberate flouting of the criminal law, which has happened, albeit very rarely.

Regulation has disadvantages too, as will be shown. But work on embryos might never have been permitted in Britain at all had it not been for the existence of, at first, voluntary professional self-regulation and, subsequently, statutory controls. Embryo research has progressed by and large in tandem with public and peer acceptability. In addressing issues relating to public fear of new technologies, family formation and safety concerns, regulation has been more of a success than a failure.

Much of the account that follows is about the profound and controversial ethical background to new developments in the field. It should not be overlooked, however, that other factors play an equal, if not more important part, in decision-making. The regulatory context itself is crucial. At first Britain was alone in the world in establishing such a system, although others subsequently followed suit.

So what is the reality of experiencing and administering regulation in the British context and are there disadvantages?

Regulation has to be financed. In the context of fertility treatment, it is the patients who bear the expense, which is passed on to them by the clinics. Relatively little IVF treatment is funded by the NHS and in those cases too the NHS has to absorb the extra cost represented by regulation, reducing the resources available elsewhere in the hospital system. The existence of robust regulation in the UK may cause some to seek to evade its constraints. The practice of reproductive tourism, that is, travelling abroad to obtain a treatment banned at home, is one instance of this.

IVF and embryology stimulate constant debate, criticism and demands for reform. There are effective lobby groups, as well as doctors and politicians always ready to oppose any decision made by the regulators. In this atmosphere, regulatory decisions come under intense scrutiny. When approaching new questions in the regulation of embryology, some issues in practice determine the outcome.

I. The legal framework of the Human Fertilisation and Embryology Act

Every regulatory decision has to be taken in the knowledge that it may be challenged in the courts, either by a disappointed individual, a clinic or by a pressure group. It is important to the regulator to succeed in the litigation. The UK has no overarching written constitution that might ensure success on the basis of a higher principle of freedom. However, its detailed regulatory legislation is boxed in or, some would say, made porous by recent human rights and EC Treaty provisions.

The application of the Human Rights Act 1998 to legislation enacted before that date has tilted interpretation towards individual rights and liberty in a legislative environment that had been carefully crafted to balance public health demands and individual desires. The EC Treaty principles of freedom of movement of goods and services, and the right to seek medical treatment abroad, may in the last analysis undo all the careful regulatory constraints applied in Britain.

II. Resources, to fight and to enforce

The regulatory authority needs to be sufficiently well financed. Litigation cannot be brought to enforce the measures of a regulatory authority unless funds are available to fight a case all the way to the House of Lords. It may be the case that litigants who have attracted popular support will be better funded by the newspapers to which they have sold their stories than the government authority in opposition.

The HFEA is occasionally chastised for mistakes that were in truth unavoidable as long as the technology is in the hands of humans and resources are finite. A well-known example was the birth of black twins to a white couple, where the wrong sperm had accidentally been used in the clinic treatment. Reprimands and demands for ever tighter regulation and inspection were made even though the HFEA lacked the resources that would make seven day a week supervision possible. This degree of rigid control would in any case be at odds with the general policy of light touch regulation.

III. The power of the media

Newspaper and television coverage is often inaccurate in representing the policy issues that have to be balanced in connection with particular cases. If there is a good human story, the more appealing of the two litigants may carry greater weight. When a young woman is featured in the media, making an appeal for a baby (by some risky method), issues of principle and precedent in the decision may be swept aside by its emotive appeal.

IV. Politics

Government departments and ministers have very particular views in relation to questions that are widely debated. Attitudes towards genetically modified food, fluoridation, reproductive technology, are all heavily politicized. The pressures the government officials exert may be indirect, but they are nonetheless forceful.

V. Ethics

Finally, if there is any room left for debate and choice in the context of the other legal and political pressures, ethical concerns are specifically addressed. Broadly speaking the HFEA has developed five ethical principles from legislation and from its deliberations over real cases day-by-day. These considerations are bolstered by widespread public consultation.

1. *The assurance of human dignity, worth and autonomy.* In line with international conventions, nobody is to be used as a convenience or as a bank of spare parts. Consent and counselling are vital. Comatose or dying people should never have gametes removed from them without their prior knowledge and consent.
2. *The welfare of the potential child.* Consideration of the need for a father is enshrined in the legislation, although this is now open to debate.
3. *Safety is given great weight.* Newly discovered treatments such as preimplantation genetic diagnosis and egg freezing have sometimes been delayed for safety checks and trials of viability. Despite public pressure and compassion for those seeking treatment, the safety of the child and mother has be considered.
4. *Respect for the status of the embryo.* Legislation lays down the parameters of permitted research and prohibits the mixing of humans and animals, cloning and research on embryos over fourteen days old.

   These four principles can be detected in decisions taken about, for example, the posthumous removal of sperm and the ban on sex selection for social reasons.
5. *The saving of life as an acceptable use to which new advances in embryology may be put.* This principle was behind the decision to extend preimplantation genetic diagnosis and HLA-typing to attempt to create a sibling whose umbilical cord blood might save an older child. It also underpins the HFEA's approach to stem cell research.

The ongoing need for regulation

Effective regulation is complex, and presents a multitude of practical difficulties. Nevertheless, comprehensive regulation is the best course wherever IVF and embryology research are being pursued.

There is a need for universal substantive legislative prohibitions in relation to cloning, and for controls on experiments in the womb and genetic manipulation. Surveillance of laboratories and clinics, and enforcement of the fourteen-day rule for keeping embryos in the laboratory are also essential. There is a case for regulation of the buying and selling of gametes, and consideration should be given to legislation banning the patenting of embryological research.

There should be universal provision for studies of the health of IVF children with publicity for the adverse consequences, if any, of certain treatments. Acknowledging the dangers of competition between clinics, there should be standards to ensure the integrity of statistics and to enable comparison between clinics. Good patient information should be provided, and a limit fixed on the number of embryos to be used in any one treatment.

Uniform safety standards should be established, and penalties for their breach. There is a need for laws on the use of embryos after the expiration of the permitted storage period, and in situations where previously given consents are unilaterally withdrawn (in cases of divorce or relationship breakdown). Patients' and donors' rights, information and consent in relation to distant research will become increasingly important and must be addressed.

An independent, central, and transparent authority which grants licences, monitors research and imposes sanctions backed by criminal penalties, is a great asset. Britain is not alone in having confidence in this method. It has been adopted in Canada, Australia, France and Japan to some degree.

The regulatory process is not universally welcomed. Scientists may resent the paperwork that stands between them and their research. Politicians may believe the HFEA holds too much power over decisions which properly belong to Parliament. Yet we live in a world where scientists are sometimes mistrusted, where there is conflict between politicians who want to control every move in this controversial area, and clinicians who want freedom to take strides into the unknown. In this environment only comprehensive regulation can hold the ring and bring order and consensus to this field.

Ruth Deech
August 2007

CHAPTER I
Reproductive Technologies and the Birth of the HFEA

Case 1: Louise Brown

Lesley and John Brown were a young couple from Bristol who by 1977 had been trying unsuccessfully for nine years to conceive a child. It seemed that they were doomed to failure, due to a blockage in one of Lesley's fallopian tubes. The presence of such a blockage means that although a woman may be able to produce eggs, they cannot travel down the fallopian tube to the uterus, and therefore cannot be fertilized inside the body.

While the Browns were endeavouring to overcome their fertility problems, two doctors were working on a procedure for removing eggs and fertilizing them outside the body. When these four people finally met, the scene was set for an experiment which would change the lives of all concerned.

Although Lesley Brown was not the first person to undergo this procedure, none of the embryos which had been implanted in other volunteers had resulted in viable pregnancies. But against the odds, the procedure did work in the case of Lesley Brown. Drs Edwards and Steptoe and Mr and Mrs Brown waited anxiously as the weeks, and then months passed. Finally, on 25 July 1978, Louise Brown was born and a new era in reproductive technology was born with her.[1]

The birth of Louise Brown sparked a flurry of media outpourings on the subject of 'test-tube babies', which still affect public opinion and the regulatory environment today. Initially, there was some dismay at the idea that a child could be deliberately created in such a way. It was feared that the baby would be horribly damaged or deformed through missing some essential early requirement from the mother's womb. Many people assumed that the child would be regarded as a second-class citizen, that she would be viewed as a pariah, discriminated against, or otherwise persecuted as a result of her origins.

The procedure of *in vitro* fertilization (IVF) was highly experimental. Ironically, Robert Edwards and Patrick Steptoe, the pioneering doctors involved, might well not have been given the go-ahead had any regulatory body been in existence at the time. Professional codes of conduct and self-regulation were soon in operation however, and in 1985 the Voluntary Licensing Authority (VLA), chaired by Mary Donaldson, was formed to oversee the application of new reproductive technologies.

Scientists wishing to use IVF techniques could obtain a licence from the VLA. This confirmed that their work was being carried out openly and was condoned by their peers. But the novelty of the procedures being developed meant that these frameworks were tested to the limits. The issue of embryo research raised further difficulties: it was unclear how such research should be regulated or, indeed, if it should be permitted at all. A feeling arose that the government should take some action to demonstrate that scientists were not simply being allowed to run amok. The public needed to see that advances in reproductive technology were being guided and monitored. With these challenges in mind, the government formed a committee chaired by the philosopher Mary Warnock (subsequently Baroness Warnock).

The Warnock Committee's task was 'to consider recent and potential developments in medicine and science related to human fertilization and embryology; to consider what policies and safeguards should be applied, including consideration of the social, ethical and legal implications of these developments; and to make recommendations'. The Warnock Report was published in 1984 (Cmnd 9314) and remains to this day one of the most influential, pragmatic and farsighted reports of its kind, seminal in the UK, and a model for other countries.

The Warnock Committee made a number of recommendations. It suggested that the new technological developments needed closer and more formalized regulation than the existing professional codes could offer. This regulation should be undertaken by a dedicated body whose decisions and workings would be transparent, and whose existence, it was hoped, would appease public fear of scientists running out of control. The Warnock Report's recommendations included, among other things, that a licensing authority should be established to regulate research and treatment services; that research on embryos should be permitted under licence; that no embryo should be kept for more than fourteen days outside the body; that fertility treatment should be based on written consent; and that embryos should be stored for a maximum of ten years.

Eventually, after lengthy Parliamentary debate, in 1990 the Human Fertilisation and Embryology Act (HFE Act) was passed, based largely on the recommendations of the Warnock Report. Its cornerstone was the creation of a statutory body to oversee and regulate IVF, donor insemination, embryo research and the storage of gametes and embryos. The Human Fertilisation and Embryology Authority (HFEA) started its work in London offices on 1 August 1991.

Since its inception, the HFEA has by turns been lauded, vilified, emulated and envied, as it has striven to keep abreast of technological advances while attempting to balance the interests of adults, children, embryos and society at large. The history of the HFEA, and the challenges it has faced, form a unique and fascinating record of the UK's efforts to construct and maintain an ethical and legal framework for the regulation of some of the most controversial and awe-inspiring processes yet developed by scientists.

We now know that predictions about the horrors of 'test-tube' babies proved largely unfounded. People saw for themselves the evidence that Louise Brown was a normal healthy child. As the IVF technique began to be used on a wider scale it became clear that IVF children did not seem to suffer particular persecution, and by the end of the twentieth century it seemed that the alarm over the first test-tube birth had been merely the shock of the new and unknown. Indeed, IVF came to be seen as wholesome and beneficial, at least where it benefited a couple who fitted into the pattern of a 'traditional' family.

However, IVF was just the start of a long procession of technological developments. In its wake came the possibility of implanting an embryo into the womb of a woman who was not the genetic mother of the child. This created a chasm between the previously inseparable concepts of genetic and gestational parenthood. The possibilities presented by freezing sperm and embryos, and the development of intracytoplasmic sperm injection (ICSI, the injection of sperm directly into an egg) widened the applications of the IVF techniques far beyond what had been envisaged just a few years earlier. Further challenges emerged with the development of preimplantation diagnostic techniques. Scientists could identify and discard embryos with specific genetic characteristics, leading to fears of a new generation of 'designer babies'. This technique in conjunction with tissue typing opened the way for renewed controversy over the creation of 'saviour siblings', where parents with a seriously ill child sought to have a healthy baby with characteristics that could help the treatment of the older child, discussed further in Chapter 3.

Then came further technological developments which again generated huge public interest. In 1997, with the birth of Dolly the sheep, it had become possible to create one mammal with a genome identical to that of another. This breakthrough was hailed variously as a miracle of modern science, or an assault on nature. Speculation about whether and how this technique could be applied to humans ran riot, often far exceeding what might ever be technically possible. Many of the concerns raised significantly overplayed the degree to which human identity is founded on genetic makeup. Nevertheless, reproductive cloning challenged assumptions about the structure of families and the concepts of dignity and reproductive autonomy, genetic diversity, and the meaning of sexual intercourse.

The development of cloning also brought into question the strength of international conventions and the diversity of practices within the European Community. The UK government, along with many nations around the world, hastened to legislate against the production of human clones, in the Human Reproductive Cloning Act 2001. In Europe, nations were divided according to their cultural and religious histories on the merits and permissibility of the various degrees of cloning. This led to the drafting of a rather vague and imprecise Europe-wide treaty—the Council of Europe Convention on Human Rights and Biomedicine 1997—which the UK government did not sign.

Just as public concern over IVF has waned, we might surmise that in the future people may become less concerned about cloning and its apparently world-shattering implications. Should anyone actually succeed in cloning a human being, and should it become relatively normal to do so, we may live to see a gradual mellowing of moral anguish over cloning during our own lifetimes. Already, the laws have been liberalized to allow for the possibility of 'therapeutic cloning'—a procedure whereby a cloned human embryo is created, not for reproductive purposes, but so that a stem cell line can be developed from it, which in turn may yield cells that can be used to treat a variety of illnesses.

The potential in embryonic stem cell therapies offers one of the greatest challenges, and perhaps the greatest hopes for the future. It also generates new problems for the regulators to consider and grapple with. It has been suggested that the development of stem cell therapies could put an end to diabetes, cancer, infertility, and even human ageing![2] The capacity of stem cell cultures to become 'immortal' (ie achieving a state in which they do not undergo normal cell death) has even led to speculation that the inevitability of death for human beings may itself one day become a thing of the past.

These possibilities, when considered in the context of past challenges, demonstrate the continuing importance of maintaining some form of regulatory control over the development and use of new technologies. Public confidence in science, technology and progress depends to some extent on the ability of bodies such as the HFEA not only to move with the times, but to project ahead, to evaluate the likelihood of these developments, and to ensure that the public interest does not get swept away either by reactionary rhetoric, or by over-optimistic promises about the future.

Before examining some of the dilemmas posed by these technological possibilities, it is worth taking a brief look at the workings of the human reproductive system, and exploring the development of the techniques designed to overcome fertility problems.

Natural reproduction: what is 'supposed' to happen, and why does it go wrong?

Natural reproduction, as everyone knows, requires a sperm and an egg to meet and fuse. Like all cells, gametes (sperm and egg cells) start out with forty-six chromosomes. Chromosomes are the means by which genetic information (DNA) is stored in a cell. However, while other cells retain their forty-six chromosomes throughout their life span, gametes undergo a complex series of metamorphoses, culminating in a process known as 'meiosis'. This involves ejecting half of their chromosomes, leaving them with only twenty-three. In reproduction, the sperm and egg each provide half of the genetic information that forms the resulting offspring. However, it is not predictable exactly *which* of the twenty-three chromosomes a parent will transmit to his or her child, since the process of meiosis also 'reshuffles' the genetic information. This is why siblings born from the same parents are not identical (unless, of course, they are identical twins). Each individual egg and sperm is genetically unique.

The process

In a fertile woman of reproductive age, an egg 'ripens' and is released from the ovary once a month. Contrary to what is popularly thought, the ovaries are not constantly brimming over with eggs ready to be fertilized; if one removed the ovaries from a woman's body, it would be unlikely that more than one egg would be functional. The ovaries are not full of eggs as such, but of follicles which could possibly produce eggs given time and the right circumstances. During the month before ovulation, one of the follicles in the ovaries begins to develop, and gradually reaches maturity. The ripe egg is eventually released from the follicle, and 'collected' by the fallopian tubes. It then travels down the fallopian tubes towards the uterus, taking several days on its journey. It can be fertilized at any time along this route.

The production of male gametes is a different story entirely, and focuses on quantity rather than quality. Sperm is manufactured in the testes, and millions of sperm cells are released in each ejaculation. If this occurs during intercourse with a woman, the sperm swim up through the cervix and into the fallopian tubes. Should this happen within a few days of the woman's ovulation, a sperm may reach and fertilize the egg. However, huge numbers of the sperm are likely to die before ever reaching the egg, because they have a limited ability to survive the naturally acidic environment of the woman's body. Despite this, the large number of sperm released improves the chance that at least some may survive for a few days, so intercourse the day before or after ovulation may also result in conception.

When sperm do meet the egg, fertilization is achieved when a single sperm is absorbed through the 'shell' of the egg. At this stage, the 'zygote' or fertilized egg, with the addition of the sperm's genetic material, contains the full complement of forty-six chromosomes. The egg then continues on its way down the fallopian tubes and into the uterus, where a soft and spongy lining has developed.

In a successful pregnancy the egg, which has undergone numerous cell divisions, and now contains about 150 cells, embeds itself into the lining of the uterus, and continues in its development. If the egg has not been fertilized, it is passed out of the woman's body together with the lining of the uterus. In other words, the woman has her period.

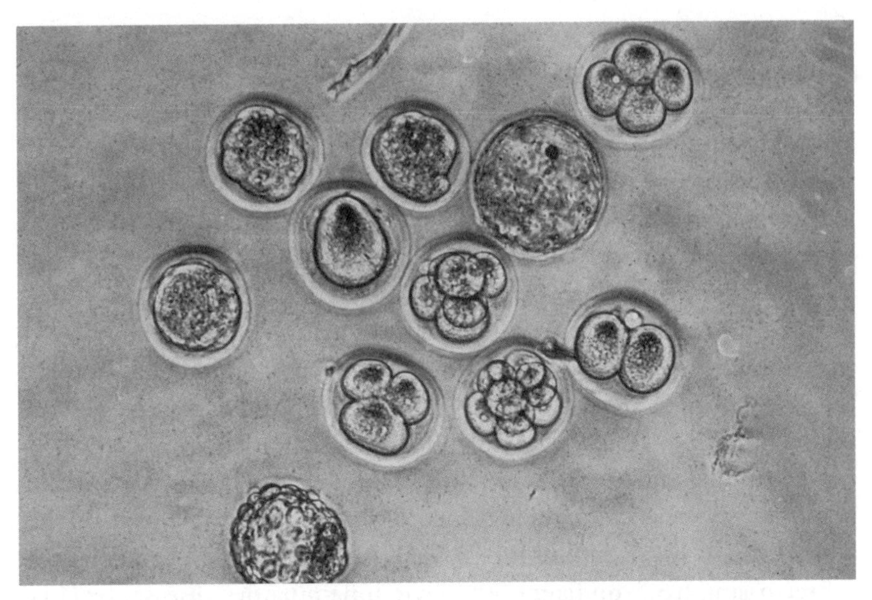

A collection of early human embryos of different stages, from the just fertilized egg to the blastocyst ready to implant in the uterine wall.

Even if conception occurs, the pregnancy is by no means guaranteed. A surprisingly large proportion (22%) of pregnancies are lost without the woman even realizing that she has conceived.[3] When figures for clinically recognized miscarriages are included, the proportion of conceptions which are lost rises to almost a third. These figures demonstrate the precarious nature of fertility and illustrate the scope for problems to arise even if a woman and her partner are fit and healthy. Nature seems to intend that not every embryo conceived should survive to pregnancy and birth. This is an interesting factor to consider in the debate about the nature—sacrosanct or not—of the fertilized egg.

When problems arise

So what can go wrong in the process of reproduction? Problems can arise at almost every stage, and may affect either or both sexes. Infertility is often presumed to be a woman's 'fault'. However, in around a third of cases where a couple is having difficulty conceiving, the problem can be traced to the man.[4] Low sperm count and problems arising from blockages in the tubes which convey sperm from the testes can affect fertility. Abnormalities in the sperm can also reduce its capacity to fertilize an egg, or its ability to swim far and fast enough to reach the egg before succumbing due to the acidic environment or the exhaustion of its energy supply. Even if the man has the capacity to produce viable sperm, psychological or physiological problems may result in impotence, with a corresponding impact on fertility.

From the woman's perspective, things are equally complex. The mechanisms and hormones involved in ovulation are extremely sensitive. Sometimes ovulation may simply not occur, and of course if this is the case, conception is impossible however plentiful and active a man's sperm might be. Being over- or underweight can affect ovulation, as can hormonal imbalances, and stress. Even if a woman is ovulating regularly, she may nevertheless encounter fertility problems if she has a blockage in her fallopian tubes. Such blockages are not uncommon (often arising from undiagnosed pelvic inflammatory disease, or chlamydia) and they may prevent the egg from reaching the uterus, as well as impeding the passage of the sperm towards the egg. A woman who is ovulating normally and has healthy fallopian tubes may have reduced fertility if there are abnormalities in the lining of the uterus. This kind of problem may result in repeated miscarriages or failure of the embryo to implant.

The problem of diagnosing causes of infertility is complicated by the fact that, in addition to the factors outlined above, many cases of difficulty in getting pregnant are simply not attributable to any identifiable underlying cause. Around one couple in every seven is known to have problems conceiving.[5] However, the real figure may be much higher than this, since it is thought that some couples may never come forward to seek treatment, and may simply accept their childlessness. (Counterbalancing this is the fact that some couples who do seek treatment may find that they confound expectations by conceiving naturally.) Modern social factors also contribute greatly to the incidence of infertility. Couples are tending to marry later in life, and relationship breakdown is common, meaning that partners may be beyond the biologically optimal age for conception by the time they are ready to have children. The spread of sometimes undetectable sexual diseases has also contributed to the rise in the numbers of those experiencing inability to reproduce when they wish to do so.

Many of these problems are associated with social rather than purely physiological factors, but this does not necessarily mean they are easy to resolve. People may be unwilling to give up their sexual liberty in order to preserve maximum fertility. Waiting for financial and emotional stability may seem preferable to women who are faced with the choice of whether to reproduce at the time when their bodies might be most suited for it, but when their lives and careers may be in a state of flux.

Before going on to look at remedies for infertility, it is necessary to point out here that not all patients seeking to use fertility treatments will be infertile per se. Some may be at risk of genetic diseases, and wish to use assisted reproductive techniques to avoid these. Others may be unable to conceive 'naturally' despite not being biologically infertile, perhaps because they are single, or because they are in same sex relationships. Fertility technologies can open the possibility of having children to people who would in the past have been excluded from reproduction by age, sexual orientation or relationship status even if they were not clinically infertile.

The appropriate treatment for patients seeking fertility treatments will depend on the cause of their complaint. IVF on its own will not work for a woman who cannot produce eggs, or for a man who cannot produce sperm. However, in conjunction with sperm and/or egg donation it may be feasible. The following section outlines some of the main techniques involved in reproductive therapies, showing how they can be used separately or in conjunction with others. Details of techniques involving embryos, including cloning and stem cell research, are also included.

Artificial insemination (AI)

Artificial insemination is perhaps the least technically complicated of assisted reproductive therapies as well as being the oldest reproductive therapy included here. AI was successfully achieved in dogs by Italian scientist Lazzaro Spallanzani in the eighteenth century,[6] and a few years later, John Hunter, a Scottish doctor, is known to have 'prescribed' the technique to a couple who consulted him with reproductive problems.[7] The patient was instructed to collect his sperm and insert it into his wife's vagina with a syringe, since physical abnormalities meant that he could not ejaculate inside her. The patient's wife did indeed have a child, although conclusive proof that this was as a result of artificial insemination has not been established.

Case 2: William Pancoast

In 1884 a case of AI using donor sperm was carried out by a Doctor William Pancoast of Jefferson Medical College in Philadelphia.[8]

A Quaker couple had approached Pancoast with fertility problems. Medical examinations revealed that in fact the problem lay with the husband's sperm, while the wife's reproductive faculties were entirely healthy. However, these facts were carefully kept secret.

Pancoast decided that this was an occasion which called for donated sperm in conjunction with artificial insemination—a procedure which had never been documented before in human beings. The woman was informed that she would need to revisit the doctor for a medical procedure designed to restore her fertility.

On arrival, she was anaesthetized using chloroform. Pancoast then selected from among his watching medical students, a sperm donor. The chosen student masturbated into a receptacle, and the semen was injected into the uterus of the unconscious woman through a rubber syringe. Her cervix was then plugged with gauze.

Nine months later she gave birth to a baby boy, still unaware of the circumstances surrounding her pregnancy.

Clearly, the relatively low-tech possibilities involved in AI can make for some ethically dubious practices. Doctors nowadays would be likely to think twice before embarking on such a deceptive course of action without patient consent. Indeed, this would be illegal. However, the fact that individuals can perform AI fairly easily means that the process is not always open to regulation. Insemination in these cases can be carried out using the 'turkey baster' technique (the sperm is sucked into a tube, inserted into the woman's vagina and then expelled from the tube by squeezing a flexible rubber bulb at one end). The HFEA has no jurisdiction over AI where it involves 'fresh' sperm and this means that companies who supply sperm over the internet, or ad hoc sperm providers are unregulated. Using sperm from unregulated providers can be risky since unscreened sperm may carry infections and/or genetic defects.

Reasons for seeking unregulated sperm donation may vary. Recent legislation dictates that children have the right to seek out their donors. However, this does not apply where sperm donation is not carried out under the auspices of the HFEA and this may be one reason why some parents prefer to seek out donors independently. Some women may prefer to obtain sperm from friends who are well known to them, and with whom they may be able to come to an arrangement about responsibility for, and access to the resulting child(ren). This kind of flexibility may seem preferable to the rigid guidelines and 'cold' atmosphere of the clinical environment for artificial insemination. However, the avoidance of such rigidity may itself bring trouble in its wake. Ad hoc family arrangements may not be recognized in law, resulting in social and practical problems for the families involved. This is especially true if misunderstandings or disputes arise.

In fertility clinics, AI may be used when a woman does not have a partner, and is using donated sperm. Alternatively, it may be used if a woman's partner is infertile or known to be the carrier of a serious genetic disease. If AI is undertaken at a clinic where sperm is being stored, this falls under HFEA regulation, and the sperm donors will be carefully vetted and screened for genetic and other health problems. Women receiving AI will be monitored to establish their ovulation times so that insemination can coincide with ovulation, maximizing the chances of pregnancy. Sometimes this is achieved with drugs but, in many cases, monitoring a woman's cycle gives a reasonable idea of when to inseminate. Once the ideal time has been established, sperm is injected directly into the uterus. Sperm used in these treatments is likely to have been stored frozen and then thawed—a process which can be achieved without damaging the sperm, or any resulting offspring.

In vitro fertilization (IVF)

Perhaps the most commonly-known assisted reproductive technique is *in vitro* fertilization (IVF). IVF is familiar to people as the process which gave rise to 'test-tube babies'. In fact the procedure is far more complex and time-consuming than this rather glib phrase implies. From start to finish, the timescale for IVF will be between four and six weeks.

IVF involves obtaining sperm and eggs and fertilizing them outside the body, rather than inside. Embryos are then replaced in a woman's uterus. But before the fertilization process can take place, it is necessary to predict and control the time at which a woman ovulates. As described above, women normally produce only one egg per month. Establishing when this egg might be surgically retrievable would be almost impossible without the use of drugs. Even if this could be achieved, given that the success rate of IVF is relatively low, it would mean that the procedure would have to be repeated many times in order to give a reasonable chance of a successful pregnancy. (Unprotected intercourse also brings only around a 20% chance of conception in one month.)[9]

Women undergoing IVF are therefore often given drugs to inhale daily in a nasal spray which actually inhibit the ovaries' production of eggs. This gives clinicians the chance to control the time at which ovulation will happen. After around two weeks of this, the process is reversed, and the next hormone, which triggers the ovaries to produce multiple eggs, is administered by daily injection, usually for about ten days. The development of the eggs is monitored by ultrasound, and tracked by measuring hormonal changes in the blood. Before the eggs are harvested, the woman is given a final injection of a drug designed to 'ripen' them. Thirty to forty hours after this dose is administered, the eggs are removed, usually under local anaesthetic and sedatives, using a fine needle passed through the vagina and into the ovaries.

Once the eggs have been retrieved, a different drug (progesterone) is introduced to prepare the uterus for pregnancy. It is administered daily either through injections, or via pessaries (vaginal suppositories), and is continued through the early pregnancy if the woman conceives successfully.

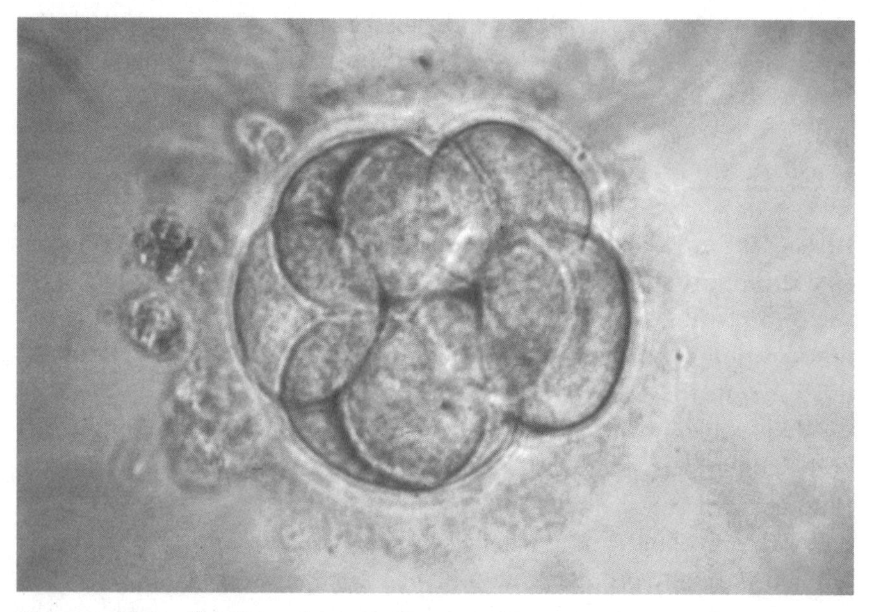

Human eight cell embryo for IVF selection. This embryo is considered to be grade one which is of sufficient quality to be used for *in vitro* fertilization. Note the evenly-sized cells, and symmetrical development.

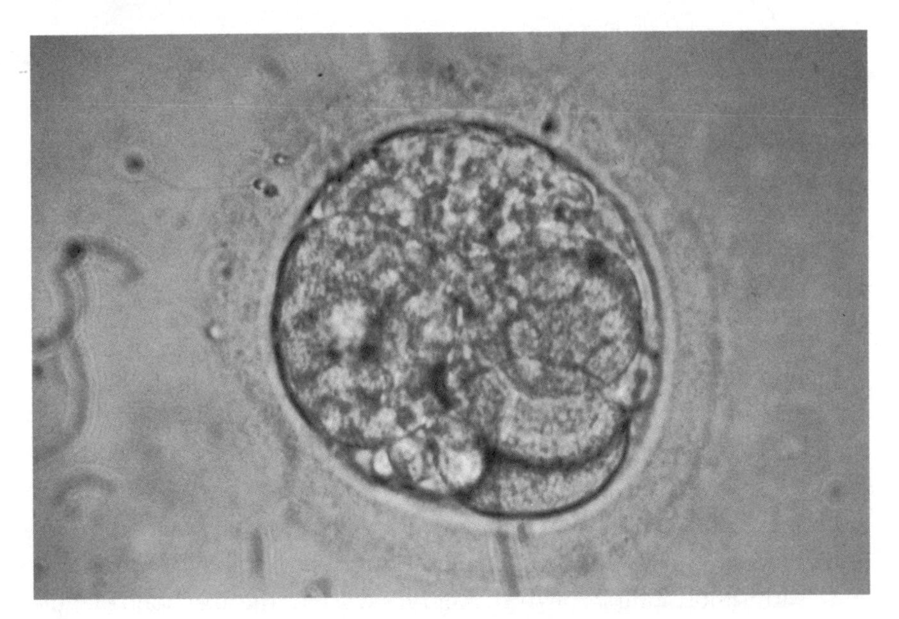

Human eight cell embryo for IVF selection. This embryo is considered to be of grade four, which is not of sufficient quality for *in vitro* fertilization. Its development is asymmetrical and haphazard.

Meanwhile, the eggs which have been collected are then mixed with her partner's sperm, usually provided by masturbation in the clinic. If a woman's partner is infertile, or if she has no male partner, then her eggs may be mixed with donated sperm. During the mixing process it is hoped that the sperm will fertilize at least some of the eggs. When fertilization has occurred, clinicians examine the fertilized eggs and allow them to begin dividing in the clinic. This ensures that any which are dividing abnormally, or which fail to divide, can be discarded. Sometimes it may transpire that no embryos have been formed, in which case, should the woman wish to keep trying, she will have to repeat the process again.

Once the 'best' looking embryos have been identified, one or two are selected for implantation into the woman's womb a few days after the eggs were first harvested. At this stage the embryos consist of around eight cells. The replacement procedure involves inserting the embryos into the uterus through a thin catheter. This is quicker and less painful than the egg retrieval (it is described as being similar to a cervical smear). No anaesthetic is usually required. For the woman and her partner, if she has one, the next phase is simply to wait and see, while continuing the progesterone treatment. Around two weeks after implantation, a pregnancy test is performed to establish whether the cycle has been successful. If so, the pregnancy and birth are expected to progress in the usual way without further interventions. If not, the woman may decide to try again using frozen embryos, or if there are none left, she may decide to embark on the drug and egg harvesting regime again.

Surplus embryos

Sometimes the process of egg collection and fertilization can result in the creation of many more embryos than can be implanted. In the UK, there is a limit of two embryos which can be transferred for any one cycle, although in patients over 40 this may occasionally be extended to three. (The HFEA has been challenged on this, as discussed in Chapter 4.) The numbers are restricted because multiple births pose serious risks for mothers and babies, and clearly, implanting six or seven embryos will raise the likelihood of multiple births. The limits are therefore designed to maximize the chances of a successful pregnancy without imposing unreasonable burdens on mothers, babies and the health service. Where there are more embryos than can safely or legally be transferred into the mother, the surplus embryos may be frozen for use in later cycles. This means that the mother will not have to go through the gruelling drug regime and egg collection again if the initial cycle fails.

Other options for surplus embryos include donating them for embryo research, or donating them for fertility treatments offered to other people.

Risks of IVF

The risks involved in IVF are necessarily asymmetrical: generally the man does not have to undergo any invasive procedure or take drugs as the woman does. However, the stresses and emotional pressures involved in IVF may affect men as well as women. Undergoing IVF can put a strain on relationships, and men may experience feelings of stress, guilt or anxiety.[10]

For women too, the tensions and uncertainties involved may take a toll. This varies hugely depending on people's outlook and experience. However, it seems safe to say that embarking on IVF is likely to involve negative emotions at some points.

In a sense, one of the greatest 'risks' attached to IVF is its relatively low chance of success. IVF can be physically, financially and emotionally burdensome, yet these burdens are often viewed as a price worth paying in order to have a child. However, in any one cycle, a woman of optimal childbearing age using her own eggs has on average only a 28.2% chance of success.[11] If the woman is older, or if she uses frozen embryos, the figures are significantly lower.[12] Undergoing the hardships of treatment provides no guarantee that a woman will come away with the child she longs for.

In addition to the likelihood of failure, there are a number of other risks which affect women undergoing IVF, whether or not treatment is successful. The drugs given to women to stimulate the ovaries can cause ovarian hyperstimulation syndrome (OHSS).[13] Where this is mild, it causes swelling and abdominal discomfort, which may be remedied by taking aspirin. In its more severe forms, OHSS can cause nausea, vomiting, sudden weight gain and fluid retention, difficulty in breathing, the formation of blood clots, and—very rarely—death. It has also been suggested that the drugs given to women to stimulate the ovaries can increase the chance of ovarian cancer. However, this association has not been supported by further studies, and it seems likely that if there is a risk, it is very low.[14]

For many women, hormonal swings and mood fluctuations will accompany the drugs used to control ovulation. The injections themselves may also cause some discomfort. Egg collection runs the usual risks of minor abdominal surgery: there is a small risk of perforating the bowel, uterus or bladder, and a possibility of infection.

Finally, the risk of having twins or triplets as a result of IVF treatment is far higher than in 'natural' pregnancy, and this in itself poses dangers to mothers and babies. The incidence of multiple births in the UK has risen considerably over the past twenty years or so, primarily because of the increase in babies born as a result of fertility treatments. Currently, 23.6% of all IVF births are twins or triplets.[15] Women who give birth to more than one child at once are at greater risk of miscarriage throughout the pregnancy, and are more likely to suffer from complications related to high blood pressure, such as pre-eclampsia. Babies also suffer. Apart from the risk of miscarriage, there is a significant risk of premature delivery, which in turn increases the likelihood of premature death, or physical and neurological damage which may affect children throughout their lives.[16]

Intracytoplasmic Sperm Injection (ICSI)

ICSI is the injection of a single sperm cell directly into an egg. It can form an additional part of the IVF process if the man has a very low sperm count, or if the mobility of his sperm is abnormal. For some men, the ejaculate contains no sperm at all due to blockages in the genital tract, previous vasectomy, or other physiological problems. In around 50% of such cases, sperm can be obtained surgically via a needle inserted through the skin of the scrotum. These men are likely to require ICSI since sperm obtained surgically are often immature and cannot fertilize an egg in the normal way. Other medical ways of retrieving sperm include removing tissue from the testicle, removal of the testes altogether, or electroejaculation, where an electric probe is inserted into the rectum, and a current passed to force ejaculation. This may result in retrograde ejaculation into the bladder, from which sperm can be removed surgically.

For the woman, the early stages of the procedure take place as described in the IVF outline above. However, instead of simply mixing the eggs with the sperm, a single sperm cell is selected by clinicians, and sucked up into a fine needle. It is then injected directly into the egg. The process is repeated to produce enough embryos for treatment, and they are then cultured and observed to ensure normal development, and implanted in the same way as normal IVF embryos.

The insertion of sperm directly into the centre of the egg in the ICSI technique bypasses the natural conception process. It is unclear whether fertilization acts as a kind of selection, weeding out defective or damaged sperm from among the many that surround the egg. There has been speculation that children born from sperm injected directly into eggs may be affected by genetic or other abnormalities. Some studies seem to bear out the possibility of health risks.[17] However, the studies are limited and further research is required to establish the exact degree of risk involved. Another possible drawback associated with ICSI is the potential for the father's fertility problems to be passed on to his offspring. ICSI does not constitute a cure for any underlying condition. This may mean that children born of ICSI have to avail themselves of the same technique in the future.

Preimplantation genetic diagnosis (PGD) and preimplantation genetic screening (PGS)

PGD involves testing embryos for genetic diseases before they are implanted in the mother. It may be undertaken when one or both prospective parents are at risk of a serious genetically transmitted disease, or when tissue typing of the embryo is desired in order to provide a tissue match for a sick brother or sister.

Because PGD entails creating embryos outside the body, many of the processes involved are the same as those involved in IVF. Eggs are collected and fertilized in the same way. However, instead of implanting the embryos directly into the woman's uterus, they are 'biopsied'. This entails the removal of a single cell from the embryo which is taken away for genetic analysis in order to determine whether the deleterious gene sequence is present. Clinicians will then select from among the unaffected embryos which ones—if any—to implant.

PGD can also be used in conjunction with human leukocyte antigen (HLA) tissue typing, where parents are seeking to have a baby who will be a tissue match for an existing sibling. This is likely to happen where the existing child needs blood or tissue donation as treatment for a serious disease,

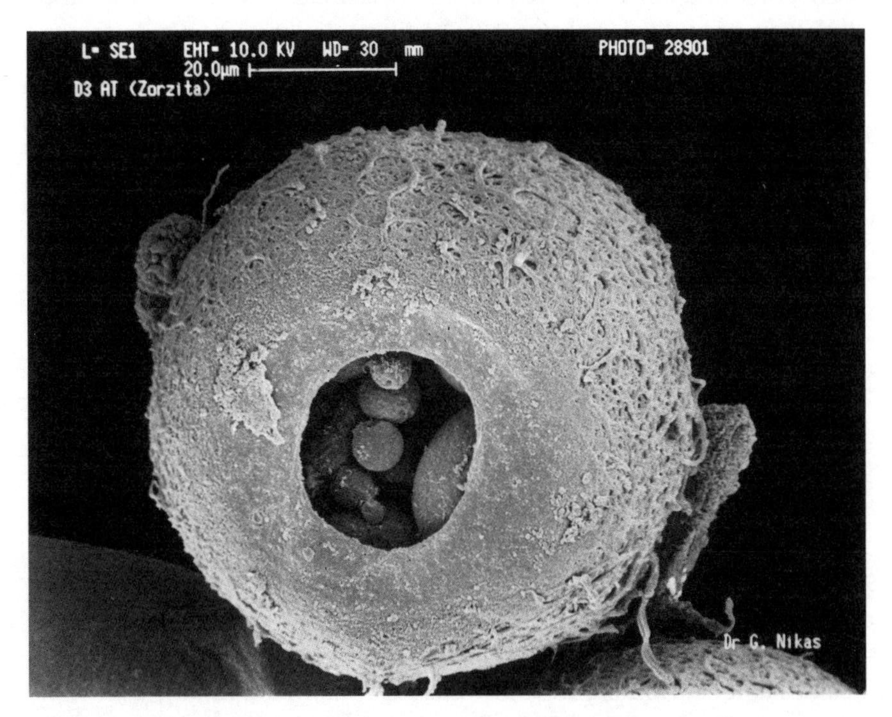

A human embryo at day 3. The egg has been fertilized *in vitro*. The coat around the egg (zona pellucida) has been treated with acid to make a hole so an individual cell can be removed. This cell can then be used for genetic diagnosis before the embryo is transferred to the woman's uterus.

and where no other matches have been found. If HLA and IVF are successful and a matching child is born, cells are removed from blood extracted from the umbilical cord just after birth, and used to treat the sick child.

Preimplantation genetic screening (PGS) works in a similar way to PGD in that it tests *in vitro* embryos following egg collection and fertilization. Where women suffer repeated, unexplained miscarriages, it is thought that this might be due to chromosomal abnormalities. Therefore it is often women in this situation who are offered PGS, especially if the patient is over 35, at which time the chances of producing genetically-abnormal eggs begin to rise sharply. Rather than looking for a specific genetic mutation, however, PGS simply counts chromosomes. A normal human embryo will have forty-six chromosomes; twenty-three from each parent. However, embryos sometimes deviate from this norm: rather than carrying two copies of each of the twenty-three chromosomes, some carry either multiple copies, or only a single copy. Chromosomal anomalies of this kind are usually lethal and are likely to be miscarried. Down Syndrome, which is caused by an extra copy of chromosome 21, is an exception to this rule.

Out of the freezer ...

Eggs, sperm and embryos can all be frozen with varying degrees of success. The processes involved in reproductive therapies often make use of freezing techniques. Donated sperm can be stored almost indefinitely, meaning that donation sessions can be kept to a minimum. (In practice, however, a limit of ten years is placed on the storage of frozen sperm.) Eggs, on the other hand, are very difficult to freeze successfully since they are much larger than other cells and contain a larger amount of fluid. This fluid can expand and develop into crystals which may damage the cell during the process of freezing and thawing. Despite this, researchers have persevered in looking for techniques which may be able to overcome these problems, and some headway has been made. In 2000 the HFEA decided to allow frozen eggs to be used in fertility treatments (it had been waiting for evidence that the technique was safe for children born as a result).

Surprisingly, perhaps, although eggs are hard to freeze successfully, embryos freeze relatively well and can be kept frozen for a number of years. This has meant that occasionally, where a couple has embryos in storage, difficulties have arisen where the relationship has broken down and one of the partners no longer wishes the embryos to be used.[18] This can be a heartbreaking situation, and one which would be avoidable if egg-freezing techniques were improved so that couples could freeze their gametes separately.

Cloning

Case 3: Dolly the sheep

In 1997 the world was stunned when news of a major scientific breakthrough hit the media. An animal had been successfully cloned. Not just any animal, either, but a sheep—a mammal—and therefore perhaps not very different from us.

Previously regarded as being the province of science fiction, suddenly the possibility of replicating human beings sprang into people's minds. For many, this was evidence that science had progressed too far. Recreating copies of other people threatened our fundamental understanding of identity and raised serious moral questions.

Under almost universal public pressure, many countries hastily issued prohibitions and decrees stating that human cloning was abhorrent, and no-one should be allowed to do it.

As time passed, the therapeutic possibilities of cloning began to present themselves, and some countries, the UK included, made a legal distinction between reproductive and therapeutic cloning. Nevertheless, international consensus on the topic is elusive. Those countries which do have legislation specifically addressing issues related to cloning, approach it in vastly different ways.[19]

A clone is an entity which is genetically identical to another entity. As we have seen, children are not clones of their parents because each parent only contributes half of their genetic makeup. Siblings are not clones of each other because each parent's genetic contribution is 'reshuffled' every time an egg or sperm is formed.

However, human clones have been in existence for as long as we have lived on this planet, in the form of identical twins.[20] Identical twins share exactly the same genetic information. How does this happen? When an egg is fertilized by a sperm, it may split and instead of developing into one embryo, may develop into two or more. Scientists are able in theory to bring about this splitting process deliberately. However, as we know, most identical twins come about for unexplained reasons, although there is thought to be a family link.

Identical twins are in fact more similar to each other genetically than engineered clones such as Dolly the sheep. This is because the egg from which Dolly came was a different egg to the one which produced the adult sheep whose cell was used to create her, whereas identical twins come from the same egg. Eggs contain genetic information which is separate from the nucleus, known as mitochondrial DNA. Therefore, although Dolly shared all her chromosomes with her genetic 'mother', there were still some residual genetic differences in this mitochondrial DNA which Dolly had received from the egg rather than from her 'parent' sheep.

For many people the idea that identical twins are clones seems to miss some of the essential meaning of the word clone, even if identical twins *are* genetically more similar to one another than Dolly the sheep was to her genetic 'mother'. It might be thought that a 'real' clone would have been deliberately created, or that he or she would have to be genetically identical to someone who was already an adult, or that he or she would have only one parent. Clearly, 'clone' is a very loaded word and may mean different things to different people.

As suggested, there is more than one way of creating a clone. One possibility is splitting embryos at an early stage when the cells are still 'totipotential', ie they still have the capacity to form either part of an embryo, or part of the placenta or amniotic sac. Another method, that used to produce Dolly the sheep, consists of the following stages:

1. An egg cell is obtained from a donor.
2. The nucleus is removed from the egg cell.
3. A 'somatic' or body cell, such as a skin cell, is obtained from the individual who is going to be cloned.
4. The nucleus from the somatic cell is extracted and inserted into the empty egg cell.
5. The egg is electrically stimulated to 'think' it has been fertilized, and begins to develop into an embryo.

At this stage, the technical possibilities diverge. Theoretically, it would be possible to allow the embryo to continue in its development and implant it into a woman in the same way as an IVF embryo. However, this is currently illegal and is likely to remain so for the foreseeable future, not

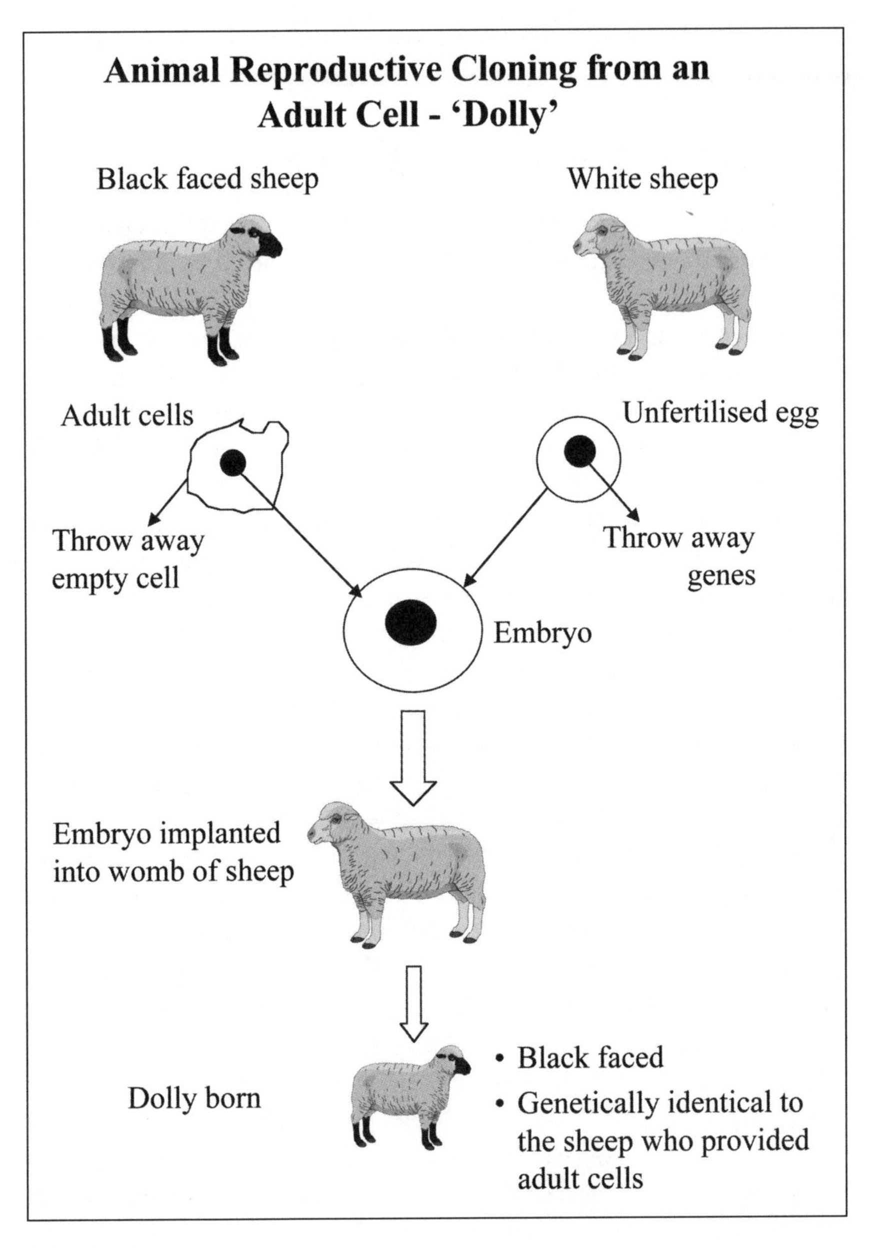
Animal Reproductive Cloning from an Adult Cell - 'Dolly'
Black faced sheep
White sheep
Adult cells
Unfertilised egg
Throw away empty cell
Throw away genes
Embryo
Embryo implanted into womb of sheep
Dolly born
• Black faced
• Genetically identical to the sheep who provided adult cells

Dolly cloning

least because the likelihood of physiological and other defects is thought to be too high.

Therapeutic cloning and stem cell research

The other possibility associated with cloning is its use in stem cell research. The aims of therapeutic and reproductive cloning are very different. One aims at creating a living replica of another entity. The other aims at creating cells genetically identical to those of a potential patient. Soon after the birth of Dolly, the HFEA realized the importance of distinguishing cloning for therapeutic purposes from cloning for reproductive purposes. The beneficial possibilities of therapeutic cloning seemed to merit legalization and regulation.

Embryonic stem cells (ESCs) are formed during the early development of the embryo. If these cells are extracted, they can be cultured, and retain the capacity to differentiate into any specific body cell, such as skin, nerve, muscle, etc. It is not fully understood how to control this differentiation process, and this is why scientists are eager to continue their research in this area. Stem cells can in theory be derived from any embryo. However, researchers are especially interested in trying to obtain them from embryos which have been formed by cloning. This is because the resulting cells would be genetically identical to the individual who had been cloned.

It is hoped that obtaining ESCs from embryos cloned from prospective patients might enable scientists to develop new cells which would exactly match the patient's DNA. This would overcome problems related to the rejection of 'foreign' cells or organs. It might also provide a way of treating diseases which attack or destroy particular cells, such as Alzheimer's disease, or diabetes.[21]

In conjunction with the Human Genetics Advisory Commission, the HFEA paper *Cloning Issues in Reproduction, Science and Medicine* (1998) recommended that a legal distinction be made between reproductive and therapeutic cloning. Legalization and Parliamentary approval followed. This initiative has been regarded as one of the HFEA's greatest achievements, for from this period the UK has led the world in both advancing and monitoring stem cell research. The basis on which it does so is that established by the HFEA in 1991 for the regulation of IVF and embryo storage.

CHAPTER 2
Ethics, Embryos, and Infertility

The excitement of pioneering has not vanished from the area of fertility treatments. Those who are involved in regulation, as well as the clinicians and researchers in the field, are aware that they are operating at the outer limits of technology in a uniquely intimate field, imbued with ethical[1] as well as scientific significance, that they are probing the depths of human origins in the face of opposition from many quarters, and that their decisions and discoveries will raise more questions, both religious and secular, and re-open old discussions. In this chapter, we look at some of the moral principles which influence decision-making in these areas. We discuss what happens when principles conflict, and address the difficult question of who can be charged with making the necessary decisions and how they should make them.

It was Dr RG Edwards, the pioneer of *in vitro* fertilization, who first articulated concerns that the ability to create and culture embryos *in vitro* would raise serious ethical problems. Seeing that these dilemmas were likely to cause tensions between scientists, politicians and the public, Edwards suggested the implementation of a 'simple organization, easily approached and consulted. It should advise and assist biologists and others, to reach their own decisions. Such an organization must represent widespread but uncommitted interests and be free of partisan politics. It would frame public debate, act as a watchdog, but interfere minimally with the independence of science'.[2]

Dr Edwards was perhaps committed to the facilitation of science rather than to the imposition of restraints based on public concerns. His ambivalent attitude toward regulation was revealed when he later described state interference in reproductive medicine as 'Nazism and Stalinism'.[3] If nothing else, this illustrates the way scientific and medical support for regulation fluctuates in relation to what the scientists may regard as the imperative for freedom in research.

At this time, another far-sighted pioneer, James Watson, was warning about the ethical problems of human cloning and surrogacy, were they ever to prove possible. Watson called for international agreement in legislating for these possibilities.[4] The HFEA was created at least in part to deal with the ethical issues associated with fertility treatments and embryo research. Prospective patients, children and embryos are the most obvious focus of concern. However, the HFEA also has broader obligations, which include protecting clinicians and scientists. The HFEA provides an environment in which clinicians can operate free from legal challenges provided that they adhere to the regulatory guidelines. Clinicians are also guaranteed a right to conscientious objection.

The HFEA's ethical challenges relate primarily to the deep-seated concerns evoked by developments in reproductive technologies and embryo research. These issues are presented in a context of intense media interest to a public that is, by and large, poorly educated in science. Public unease over developments in reproductive technologies may indicate deep moral disgust; equally, it may simply represent a fear of the unknown.

Morality is intricately connected with cultural or religious perspectives. But it can also be approached through reason and objective analysis. Either way, there is ample scope for disagreement. In the UK, the one thing of which we can be certain is that diverse moral views will exist. It is this diversity of opinion and belief that poses such a challenge to an institution such as the HFEA, whose judgements must be comprehensible—if not always welcome—to the public at large.

Relief of infertility as a worthy goal

It is often assumed that relieving the harm caused by infertility is in itself good. However, this is an assumption which could be further unpicked. Being infertile is not necessarily associated with physical risk, or adverse health outcomes. Should we feel uneasy about the welfare of women who undergo painful and sometimes dangerous fertility treatments? Perhaps those who find that they are infertile should be encouraged to accept their lot and seek fulfilment in other areas of their life.

A joint report published by the Human Genetics Advisory Commission and the HFEA stated:

The relief of the pain of infertility is, in general, a good end, but it is not an absolute end to be achieved without regard to the ethical acceptability of the means employed for that achievement. The wish for genetic offspring is a natural human aspiration, but this has to be held in balance with other desirable aspects of human wellbeing and it cannot be given an overriding priority above all other considerations.[5]

Even if providing treatment options for the infertile is a worthy aim, it does not necessarily eclipse other moral concerns. Moreover, the rapid development of reproductive technologies means that there is often scope for doing more than simply relieving infertility. The ways in which techniques are put to use can extend the choices open to prospective parents, sometimes in unforeseen ways. In this rapidly changing environment moral concerns are never far away.

Case 4: Bespoke embryos

In 2006, the *Daily Mail* published a news story describing a new trend in fertility treatments.[6] A clinic in America was offering paying customers the chance to choose embryos to match their specific requirements. Generally, when a couple 'adopts' an embryo, it is donated to them by a patient who has undergone fertility treatment and has surplus embryos left in the clinic. Thus, its genetic characteristics are already fixed. However, the embryos at the American clinic were being created especially to order.

The clinic had on its books a number of egg and sperm donors (who were also paid for their services). This meant that it was relatively simple to ask their infertile clients to specify the kind of characteristics they would like their embryo to have. The clinic could then consult its donor list and select the two 'parents' who would most closely match their clients' requirements. The sperm and eggs would be mixed accordingly, and the resulting embryo 'adopted' by the clients.

This example shows the disturbing possibilities raised by facilitating consumer choice in the context of reproductive technology. But the article is also fascinating because of the varied responses it elicited. Readers were invited to express their views on the story online. A glance at these comments[7] shows that there was a vast range of different views. Because embryos were effectively being bought and sold, some commentators thought the practice was akin to slavery. Others saw no harm in trying to select the characteristics of one's future offspring.

It is perhaps inevitable that in the face of such a broad moral spectrum, decisions made by the HFEA will be viewed as unethical by at least some people. Because of this, the HFEA places emphasis on making its reasoning clear to the public, and openly acknowledging the principles on which its judgements are based. And since technology is continually moving forward, and the social environment is always changing, the need for regular review is essential. This chapter gives an account of the principles on which the HFEA bases its judgements. It goes further to explore some of the assumptions on which these principles are based, and points out the difficulties of achieving consensus.

Ethics, law, and regulation

Complex ethical questions have to be addressed in the context of what is permitted or proscribed by law. The Human Fertilisation and Embryology Act is a valuable piece of legislation in the context of reproductive questions and embryo research. Other relevant legislation includes:

- the Human Rights Act 1998;
- recent European Conventions on Biomedicine and allied matters; and
- provisions in the Treaty of Rome, such as the freedom of movement to seek medical services and the freedom of movement of goods.

There is also a body of useful international guidance drafted by non-governmental organizations, including:

- the World Medical Association Declaration of Helsinki on Ethical Medical Research; and
- the statement of the Ethics Committee of the Human Genome Project.

The two most influential statements of principle on the international scene are:

- the UNESCO Universal Declaration on the Human Genome and Human Rights (1997); and
- the Council of Europe's Convention on Human Rights and Biomedicine (1997).

In a number of countries, official bioethics advisory bodies have addressed the new genetics, for example, the National Bioethics Committees of France and Italy. In Britain the Human Genetics Commission performs this non-executive function.

Most of the ethical considerations that underpin these frameworks are focused on safeguarding the individual. But it is recognized that the public good ought to be promoted too, for without advancing benefits for society at large, there is little scope for improving the lot of individuals. To quote the Human Genetics Commission: 'it is important to see the individual as a member of society with a shared interest in medical progress and the conquest of illness ...'[8] Tensions between individual and social interests are at the heart of many of the dilemmas explored later in this book.

The relationship between ethics and law is itself the subject of much debate. While moral concerns are an essential aspect of legal and regulatory frameworks, the law cannot undertake to reflect the diversity of beliefs held by the public. To complicate the issue, the law itself may influence moral feeling. Legislation passed to permit embryo research, for example, may encourage the public to view such research as being acceptable. The Warnock Committee confronted this issue in its consideration of whether, and how, to regulate embryo research. Legalizing embryo research may not have swayed its most vociferous opponents but it may have persuaded some of those who were merely doubtful.

People may hold utterly conflicting moral views, and may hold them passionately and sincerely. How can we deal with this fact? While it would be useful to have recourse to some objective measure of morality, we simply cannot *know* which of two opposing views is morally correct in the way that we can establish which of two factual claims is correct. So why should we favour certain moral claims over others?

If society were unified in religious belief, this might be helpful in terms of providing a coherent framework for morality. We might then find that enshrining the moral principles of religion in law would provide a consistent and acceptable way of proscribing certain behaviours.[9] (It is worth observing, however, that even where apparently there is homogeneity in religious belief, there is always scope for theological differences in interpretation.)

A universally accepted system of morals derived from a common religion and enforced by law is not feasible in the UK today. This means that the way in which we evaluate moral claims, and how we choose to implement them in law is an intricate and controversial affair. One of the most obvious ways of incorporating ethics into legal and regulatory frameworks is to take account of the majority view on a particular issue. This way if most people feel that, for example, embryo research is acceptable, the law can legislate to allow for such research to take place, and if not, it can be prohibited.

Yet the idea that morality is simply a question of the majority view is not appealing. In the past a majority might have found it acceptable to keep slaves or to deny women the vote. But today many of us hold these things to be wrong not simply because the majority view has changed, but because we believe that in the past, the majority were mistaken. This is an unnerving idea, since it implies that what seems so very right today may also be seen as plain wrong in one hundred years' time.

The dilemmas and ethical problems addressed here are enmeshed with questions which have vexed philosophers for centuries. We cannot provide definitive answers here. However, we can at least explore some of the complexities. We can also try to establish what considerations should be addressed in trying to base regulatory frameworks on sound ethical principles.

Respect for autonomy

Respect for autonomy is a central tenet of medical ethics in the UK and many other countries. Indeed, some regard it as a universal ethical principle which can be recognized and accepted by members of any religion or culture.[10] Autonomy can be described as self-government: an autonomous person is someone who exercises the power of making his or her own choices and decisions. In the medical context this means that doctors do not have the power to force us to make particular treatment decisions, even if our own choices are clearly harmful to us.

The ethical importance of autonomy is often associated with the German philosopher Immanuel Kant, who famously said that people should never be treated as mere means, but should always be treated as an end in themselves. (Ironically perhaps, the most infamous example of people being treated as mere means on an industrial scale occurred in Germany 130 years after Kant's death. It may be these events themselves that have so starkly reminded us of the importance of Kant's injunction.)

Kant's dictum has been the subject of much speculation (for example, what does it *mean* to treat someone as a mere means?), but its intuitive appeal is undeniable. Often, where people have a conviction that something is morally wrong but find it difficult to articulate why, this principle may shed light on the matter. However, while respect for autonomy may be appealing, it is not sufficient to build up a whole moral framework based on this principle alone. More specifically, it is not feasible or desirable to sanction all people's actions in law on the grounds that we must respect their autonomy. Some constraints are necessary.

The philosopher John Stuart Mill claimed that:

> ...the only purpose for which power can be rightfully exercised over any member of a civilized community, against his will, is to prevent harm to others. The only part of the conduct of any one, of which he is amenable to society, is that which concerns others. In the part which merely concerns himself, his independence is, of right, absolute. Over himself, over his own body and mind, the individual is sovereign.[11]

Mill also places high importance on respect for autonomy. However, his injunction provides clear instructions as to the circumstances in which autonomy can be overridden. People in general are not to be prevented from going about their enterprises unless it can be shown that harm to others will ensue. If we accept Mill's argument, the default position would be to allow scientists to perform research, or to allow patients to access treatment, as a part of the exercise of their autonomy. However, where harm to others (the newly created baby, for example) does follow from this, then freedom may be restricted, whether by prohibitive legislation or the denial of a licence. So enterprises such as reproductive cloning are illegal partially at least because they are thought likely to be harmful to the cloned person. (Questions as to whether embryos are harmed during the course of fertility treatment or embryo research are addressed in Chapter 8.)

Autonomy and safety

Case 5: Mandy Allwood

In 1996 a major news story hit the headlines. Mandy Allwood had been receiving fertility drugs. She had been informed that there was a risk of producing multiple eggs, and that she should use contraceptives. However, she had failed to comply, and was now pregnant with octuplets.

Ms Allwood had been strongly advised to undergo selective abortion in order to maximize the chances of at least some of the foetuses being born alive. However, she was adamant that she would not terminate any of the octuplets, and would take the chance that they would survive the pregnancy. Doctors were highly sceptical that this course of action would result in any live births.

It was subsequently reported[12] that Ms Allwood had made a lucrative deal with a newspaper. As well as paying for exclusive interviews, the *News of the World* would pay a cash sum for each child born alive. Did this give Ms Allwood an incentive to refuse selective abortion...?

The outcome was tragic. All of the octuplets were lost nineteen weeks into the pregnancy.

Fertility treatments involve an intricate nexus of entities whose moral relationships are all intertwined. Patients, embryos, babies, families, clinicians, and society must all be taken into account. The case of Mandy Allwood illustrates the difficulties involved in trying to safeguard these various interests. Many people may have thought Ms Allwood culpable in choosing not to 'reduce' her pregnancy. Some might even have felt that her autonomy should have been overridden and that she should have been forced to undergo selective abortion. However, others felt that reduction would have been worse than allowing the pregnancy to run its course. There is always a possibility of harm in the context of fertility treatment—new technologies may carry unforeseen risks, or clinical advice may be ignored or misunderstood by patients. When unexpected events occur, the avoidance of harm must be balanced against the autonomy of patients and clinicians to pursue their own aims. This is no easy task.

Respect for autonomy is a major component of the ethical concerns we might have for adults undergoing fertility treatments. But in some circumstances it seems to conflict with other important moral principles. The HFEA is involved in making decisions on many new and groundbreaking issues. It is not always easy to strike a balance between avoiding harm to prospective patients, and allowing individuals to pursue new ways of achieving their desires. In examining the case for each proposed new technique the HFEA will, through one of its expert groups, and with external advice, wish to establish that the applicant has shown that:

- the technique is safe;
- its reliability has been demonstrated;
- adequate pre-clinical research has been conducted;
- the applicant has the appropriate skills and experience to perform the technique;
- appropriate patient selection criteria have been established; and
- appropriate information and consent procedures are in place.

There are limits in fertility treatments, as in all medicine, as to what risks can reasonably be assumed by an individual, however well informed. Some adults are not only willing, but eager, to seek treatment even in the face of considerable risk. These individuals may not be impressed by the idea that they need to be protected: surely they should be free to make their own decisions about the risks they wish to take? To some extent, any regulation of fertility treatments might be seen as paternalistic insofar as it restricts people's freedoms on the grounds of protecting them from dangers which they might otherwise be willing to undergo.

The statement from Mill cited above suggests that an individual's freedom should be restricted only to prevent harm to *others*, not to prevent harm to that particular individual him- or herself. Generally, in our society, individuals are regarded as having the right to make decisions which may be harmful to themselves, and these decisions are respected. This means that, for example, patients are not forced to receive life-saving treatments if they do not want them. This might encourage patients receiving fertility treatments to argue against being prevented from pursuing their reproductive ends: risks to their own wellbeing, they might claim, are their own affair. Indeed, this tension between individuals' willingness to undergo risk and the role of the HFEA in preventing harm to patients is a fraught area.

In fact, although respect for autonomy seems to dictate that patients should not be forced to undergo treatments, it does not work the other way round. A patient cannot demand a treatment which would inevitably result in her death even though he or she can refuse potentially life-saving treatment where this refusal will result in death. At first glimpse, this may seem contradictory. If the end result is the same, why make what seems to be an arbitrary distinction between the two?

One answer to this is that when a patient refuses life-saving treatment, there is no specific person—other than the patient—who is obviously responsible for the perpetration of that harm. But if the patient demands harmful treatment, this becomes the equivalent of requiring of the doctors that they inflict harm upon him or her. This might well be regarded as a demand which goes beyond what is reasonable. While we might think that respect for our autonomy gives us the right to harm ourselves, it would be dubious to infer that it also gives us the right to demand that others harm us.[13]

In fact, many patients welcome the HFEA's role in ensuring safety, even at the cost of some degree of liberty. Without these safeguards, the onus would be on the individual to investigate in order to try to distinguish reputable clinics and reasonably safe treatments from unscrupulous practitioners and unproven procedures. Medical treatments in general have to be rigorously tested and proven safe before they are allowed to be used on patients. In this respect the HFEA is in line with more general attitudes towards the question of risk in medical treatment, although the wider issue is still debatable: should patients be denied treatments which may harm them?

Autonomy and the public good

In the context of fertility treatment, conflicts between patient autonomy and possible harm to the patient are usually complicated by other factors. The issue of harm to others also plays a part. The risks of experimental or unproven treatment affect not only the individual who undergoes treatment, but also the rest of society. Unsafe fertility treatments may impair a woman's health and ultimately lead to a need for further medical intervention. They may also result in the birth of damaged children with additional healthcare needs that may persist throughout their lives. Even if patients pay privately for their treatment, the economic consequences of their choices may rebound on public funds. This means that there is a wider social interest in regulating the kinds of treatment available to UK citizens.

The implantation of multiple embryos is a good example of this kind of risky treatment. Clinicians and patients are understandably eager to maximize the chances of a successful pregnancy. This may mean that it is tempting to place as many embryos as possible in the patient in the hope that at least one will successfully implant. However, this hugely raises the risk of multiple births. This is a serious problem, since the risks to mother and children from multiple births are far higher than those associated with singletons. Any additional burdens and expense of multiple births will fall in the end to the NHS and social security.

So while respect for patient autonomy, even at the cost of that individual's own wellbeing, is a requirement of extremely high importance, it may be overridden by the economic and social consequences of risky treatment. Following up these consequences may be difficult, but it remains an important consideration.

Autonomy and consent

One principle is almost universally accepted in regard to experimentation on volunteers, new medical treatments and work involving human gametes. This principle is that subjects should have full information and be enabled to give their free, withdrawable, consent. After all, individuals cannot act autonomously unless they have accurate information on which to base their decisions. Because of this, informed consent has become a vital prerequisite for any medical treatment.

This may be challenging, since treatments and the science behind them can be extremely complex. Scientists and clinicians may struggle to convey accurate information to patients who do not share their expertise. Explanations given in emotive or persuasive language may also compromise a patient's ability to make autonomous treatment decisions. Fertility treatment comes with no guarantee of success, only statistics and probabilities. These are notoriously hard to communicate in ways which can be readily understood by ordinary members of the public.

If communicating complex statistics and scientific procedures is difficult, achieving this in an unbiased, non-directive way may be still harder. Undergoing fertility treatment can evoke deep emotional and psychological responses. This may leave patients feeling vulnerable and more suggestible than usual. Inevitably, clinicians will have their own views, but they need to ensure that their views do not sway patients unduly. One of the means by which this is avoided is by ensuring that all patients have access to non-directive counselling—a stipulation of the HFE Act.

The relatively low success rates for IVF treatment must be made clear to patients. Where treatment is being paid for privately, it is also important that clinics are clear about the charges involved, and do not spring surprises on patients who have already committed themselves to a course of treatment. For example, some patients have found themselves being charged for drugs which they had assumed were included in the overall cost of treatment.

The question of informed consent was an issue from the earliest days of IVF. Steptoe and Edwards had been accused of being cavalier in this respect.[14] It became clear that future regulation would have to make very stringent provisions to ensure that patients fully understood what their treatment entailed, and that they were not coerced in any way. For this reason, the HFE Act specified that consent must be given in writing before any use of gametes or embryos.

If giving consent to medical procedures is difficult, it is doubly so in the context of fertility treatment, and sometimes quadruply. British law has attempted to deal with all foreseeable circumstances by requirements of consent in the HFE Act. Yet despite these precautions things can still go wrong.

Case 6: Mr and Mrs A and Mr and Mrs B[15]

Mrs and Mr A and Mr and Mrs B were two couples experiencing fertility problems. Both couples attended a clinic in order to receive treatment, and both were advised that ICSI might be the best way to remedy their problems.

Accordingly, the wife in each case had eggs removed, while the two husbands donated sperm for the fertilization of their wives' eggs. Each party gave written consent for their gametes to be fertilized with those of their spouse. No other use of the gametes was permitted.

After implantation, Mrs A became pregnant. She gave birth to twins, but it soon became apparent that something strange had occurred. Mrs and Mr A are both white, yet their children were visibly of mixed race.

It transpired that the clinic had injected the sperm of Mr B into Mrs A's egg, and thus the mystery was resolved. But with this new information a morass of legal and ethical difficulties emerged. None of the four adults had consented to the mixing of their gametes with an individual other than their spouse.

After much deliberation, it was determined that Mr B was the legal father of the twins, but that the children would remain with Mrs and Mr A.

This kind of mix-up raises some perplexing questions about the legal, social and biological boundaries of parenthood. It also shows that however painstakingly consent protocols are drawn up, there may still be scope for problems to emerge. In this case, it was the result of an error by the clinic rather than a flaw in the consent process itself. But difficulties around the validity of consent can also emerge when relationships break up. Legal and moral wrangles develop as ex-partners re-evaluate their previous reproductive choices. If a couple splits up, either party can withdraw consent to the storage or use of embryos that have been created during the course of the relationship. This can cause terrible distress for those whose only chance of having a child may be thwarted by an ex-partner. (A case related to this is discussed in depth in Chapter 4.)

The consent provisions of the HFE Act are designed to circumvent this kind of problem by specifying, as far as possible, every eventuality. This is one reason why ongoing consent is required, rather than consent given at the time of treatment or storage simply being regarded as binding over time. Patients consenting to the storage of gametes or embryos must also specify the length of the storage period (within the legal ten-year maximum period). Patients must also state what is to be done with the gametes or embryos if either partner dies or becomes incapacitated.

Fertility treatment often involves the creation of surplus embryos, so people who have undergone IVF also have to decide what to do with the 'spares' after they have completed their families. Unwanted embryos may be donated to other couples, kept in storage, or donated for research. Again, specific consent is required for these decisions. Yet becoming a parent is one of the most life-changing events one can experience. Because of this, people often find that their assumptions or values change. In these circumstances, consent given prior to treatment for the disposition of embryos or gametes may no longer seem valid.[16]

For adults in this situation, leaving their embryos unclaimed in clinics may be preferable to the idea that 'their' child would go into the world in circumstances beyond their control on being donated to another couple. Donating embryos for research is also a difficult choice. It is perhaps not surprising that potential donors, feeling caught between two unappealing options, sometimes disappear from clinics' records, leaving their spare embryos unclaimed.

In 1996, this problem came to the fore when a large number of embryos created in 1991 reached the end of the five-year storage period consented to by their progenitors. Many of the former IVF patients simply could not be traced, leaving the embryos in a legal limbo. Further storage was illegal without specific consent, as was donation to research or to other couples. Embryos abandoned in this way were allowed to perish. There is something very sad about this when so many people desperately long for children.

In these cases, as in most aspects of fertility treatment, parental consent has prevalence over any moral interests which the embryos might be thought to have (eg to be 'adopted' by another patient) or any interests that scientists or society at large might have. This is something which may in itself be questionable.

Autonomy and privacy

Reproduction is often seen as a deeply private and personal endeavour. Adults do not generally have to seek permission to have children, nor are they required to specify their reasons, or to provide evidence of their parenting skills. If prospective parents' choices and decisions become public, they may feel constrained directly or indirectly by the opinions of others. Going to a clinic and discussing one's fertility and treatment options with clinicians and counsellors can leave patients feeling that the most intimate parts of their lives are a matter of public knowledge. Problems can also arise for women whose working lives are disrupted because of the demands of treatment. Employers and colleagues may speculate about sporadic absences.

Some of these problems are beyond the jurisdiction of clinics or the HFEA. However, it is important to ensure that patients' privacy is respected during the course of treatment. Stringent protocols to ensure that data are kept confidential are part of the normal good practice of any clinic, but it can be difficult to reconcile patient privacy with the wider question over the safety of some of these techniques. Patients who use IVF or ICSI or other fertility treatments may not wish others—possibly even their GPs—to know of it. Yet if the safety and long-term consequences of these treatments are to be satisfactorily monitored, follow up is essential.

At the time of writing, the oldest of the offspring born as a result of IVF, ICSI, PGD and other technologies are still young adults, and so the possibility of long-term health problems cannot be ruled out. This means that a balancing act must be achieved between respecting the privacy of patients, and ensuring that the public need for reassurance of the safety of these techniques is also satisfied.

Another problem related to parents' need for privacy is the requirement of the HFEA Code of Practice (7th edn) that clinics must 'take into account the welfare of any child who may be born'.[17] Clinics must also 'take all reasonable steps to verify the identity of those seeking treatment',[18] usually by contacting the patient's GP. In previous editions of this Code, the GP was to be specifically consulted as to the suitability of the patient to be a parent. Increasing numbers of GPs would not answer enquiries from clinics. Some refused on the grounds that it was a breach of their patients' confidentiality. Others felt that they could not judge any person's fitness to be a parent, given that those who are not infertile do not have to undergo any such scrutiny.

The requirements have been scaled down, but they may still be felt to be intrusive to patients who might not wish to have details of previous relationships, or indeed their 'fitness to be a parent' probed into. Likewise, the involvement of the GP may be seen as an additional intrusion. But given the importance of child protection within our society, and the significant reaction of the public to cases of child suffering, the idea that such safeguards should be removed remains a matter for debate.

Parental autonomy and harm to children

Harm to children born as a result of treatment is one of the most obvious concerns relating to reproductive therapies. Often if treatment seems likely to harm the patient, the prospective child is also likely to be at risk of harm. While respect for autonomy may dictate that in certain circumstances an adult can assume risks on his or her own behalf against a clinician's better judgement, it does not entail that an adult can impose risks on other individuals.

Children who may be born as a result of fertility treatment cannot, of course, be consulted as to how much risk they are willing to assume. It would therefore seem best to avoid risk of harm as far as possible. The matter is complicated by the possibility that in some circumstances the birth of one child may prove significantly beneficial or even life-saving for another child, as seen in so-called 'saviour sibling' cases where PGD is used in combination with HLA testing.

A unique requirement of section 13(5) of the HFE Act concerns the welfare of the child born as a result of assisted reproductive services.

Doctors are required by law to take account of the welfare of the child (including the child's need for a father), and of the welfare of other children who may be affected, before deciding whether to provide treatment. Attempting to define the welfare of the child is not an easy thing to do. When the law was passed by Parliament, the Lord Chancellor amplified this by saying that material considerations are important, but other things are more so: stability, security, loving care and understanding, and the warmth and compassion essential for the full development of the child's character and talents.

There are a number of ways in which reproductive technologies impact on our assumptions about children's welfare. Firstly, and most obviously, perhaps, there is the issue of health. It seems clear that it would be wrong to allow the use of procedures which are likely to compromise the health of children. The second issue relates to the ways in which the development of new techniques challenge our notions about what constitutes a parent, or a family. We will consider these two issues in turn.

Health risks to children

Case 7: Carolyn Neill[19]

In 1999, Carolyn Neill, a 34-year-old woman from Belfast, was diagnosed with cancer, and told that the radiotherapy treatment would, in all probability, leave her infertile. Ms Neill had not had a family, and at the time of her treatment, she did not have a partner. Therefore, she decided to have some of her eggs frozen, in the hope that once she had won the battle against cancer, she would still have the chance of having a family.

Ms Neill underwent the egg-freezing procedure and then embarked on the cancer treatment, which was successful. After having fought off the cancer, she decided that the time had come to start a family with her frozen eggs. To her dismay, however, she was told that while freezing eggs was legal in this country, doctors were not allowed to fertilize them, or even to thaw them.

This seemingly illogical position infuriated Ms Neill, who felt that it was inconsistent to allow eggs to be frozen if they could not be used. The HFEA held that the use of thawed eggs had not been proven in terms of safety or efficacy, and therefore their use should remain illegal. While the use of frozen eggs was permitted in other countries, success rates were as low as 1%, and the likelihood of adverse effects on children born from frozen eggs had not been established.

Ms Neill was not satisfied with the HFEA's defense of the ruling, and the HFEA was faced with the possibility of human rights litigation. Reluctantly, the ban on using frozen eggs was lifted.

More recent research has shown no evidence that children born from frozen eggs are at risk of damage. However, the success rate remains extremely low.

The HFEA's initial refusal to grant Ms Neill permission to thaw her eggs, or to take them abroad, was severely censured by some of the media. The stance adopted by the HFEA was designed to avoid the danger of babies being born with serious defects. But this was understandably felt to be frustrating to the woman who wanted to use the procedure.

Ground-breaking techniques in assisted reproductive technology (ART) all have an element of uncertainty about them. We cannot ever be entirely sure of any effects on the first child to be conceived in a test tube, or to have undergone a biopsy at the embryo stage, or to be born from eggs which have been frozen and then thawed. All of these procedures involve a degree of experimentation. Even if substantial animal testing has been carried out, there is the chance that something unique about the human species may result in unforeseen adverse effects. (The events at Northwick Park in 2006 seem to bear this out where, despite extensive animal testing, human volunteers undergoing clinical trials suffered life-threatening reactions.)

Generally, the perceived danger of abnormal babies arising from the use of IVF or other reproductive technologies has not been borne out by reality. Caution has been exercised, and where risks have been feared, certain techniques and treatments have not been made available. Sometimes this can cause grief to individuals whose wishes are thwarted. It may be doubly upsetting where this caution has eventually been shown to be unnecessary, as in the case of Ms Neill's frozen eggs.

The question we need to ask ourselves is: should some treatments be refused on the grounds that they would be harmful, or too risky to the child? For most people, it seems obvious that the answer is 'yes'. If we knew in advance that a procedure would cause a child to be born with spina bifida, it would seem wrong to allow that procedure to be used. But there is a whole wealth of difficulty behind this apparently straightforward example. For example, if there was a one in 100,000 chance of the treatment resulting in a child suffering from spina bifida, this might seem acceptable. In addition, the severity of suffering associated with a condition must also be considered. But this is not always easy.

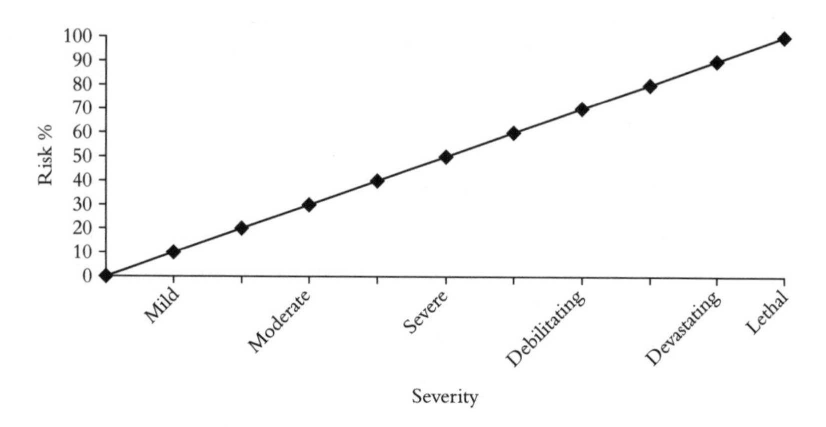

Imagine a graph whose y axis is the degree of risk and whose x axis is the severity of suffering associated with a particular condition. Can we really identify a position on that graph beyond which treatment is unethical? Where would Huntington's disease appear relative to Duchenne Muscular Dystrophy? While we are generally happy to use our unquestioned assumptions about risk and suffering to make broad moral judgements, it becomes uncomfortable if we are expected to quantify these variables. Yet if we use these concepts in our moral judgements without being willing to quantify them, can we really be said to be relying on them in any kind of relevant sense at all?

Part of the problem here is connected with the concept of suffering itself. It is easy for concerns about the suffering of offspring to become entangled with other issues. For example, many parents might prefer not to have a child with Down Syndrome. Yet this is not necessarily because such children suffer because of their condition. Some people with this condition live happy and contented lives. If parents do not wish to give birth to a child with Down Syndrome, it may be connected with their hopes and aspirations for the child and the impact on their own lives rather than because of the suffering of the child itself.

There are two important points which may be made here. Firstly, people with Down Syndrome may suffer from implicit discrimination or overt harassment and teasing even if they are not harmed by factors directly associated with their condition. Secondly, it could be argued that people with Down Syndrome do suffer insofar as they lack the range of capacities of a 'normal' human being.

Being discriminated against or stigmatized can undoubtedly cause suffering. However, there is perhaps an intuitive feeling that if people suffer as a result of social injustice, it is the injustice itself which should be dealt with, rather than attempting to prevent those who would be unjustly treated from being born.

The lack of full human capacity raises an interesting question. If someone has a condition which does not cause suffering, but does involve the lack of a particular capacity, can they be said to have been harmed? It is difficult, if not impossible to answer this question. However, there are important considerations to bear in mind. Some conditions make it unlikely that the child will ever be able to live an independent life. This might not cause suffering per se, but the capacity to be independent—to be autonomous—is extremely important to most of us. Certainly the loss of independence suffered by people who have had strokes, or serious illness, is felt acutely, yet if one has never had this independence, one cannot suffer its loss as such.

There is, however, a broader value to be associated with independence. People who cannot look after themselves impose greater costs both on their families and on the state. Families may gladly accept these burdens, but they take a toll nevertheless. Moreover, those who are dependent on others are terribly vulnerable. Undercover reports from care homes and hospitals regularly highlight exploitation and abuse of those who are unable to protect themselves. All other things being equal, then, perhaps it is reasonable to prefer that, whenever possible, people are born with full mental capacity. This could be seen as a legitimate concern for the benefit of society at large as well as a concern for the wellbeing of those born as a result of fertility treatment.

Family setup and the welfare of the child

So far we have been looking at the possibility of what might be termed 'clinical harm' to children born of reproductive technologies. But as the case below shows, there are also concerns over 'social harm'. Reproductive technologies provide scope for creating new forms of family. Children can be conceived long after their genetic parents have died. Are children harmed by being brought into the world in these circumstances?

Case 8: Mr and Mrs U[20]

Mrs and Mrs U had married in 1993. For both, it was their second marriage. Mrs U had no children, but Mr U had fathered two children who were now teenagers, and had undergone a vasectomy.

Mr and Mrs U wanted to start a family. An attempt was made to reverse his vasectomy, but this failed. The couple then considered the possibility of using donated sperm. However, they became aware that it might be possible to retrieve sperm from Mr U surgically. The sperm could then be used to fertilize Mrs U's eggs by ICSI.

The sperm was duly retrieved. Mr U signed a consent form which specified that if he died or became incapacitated, his wife could still use the sperm. However, the particular clinic at which they were receiving treatment had a policy against the posthumous use of sperm, believing that it was unethical deliberately to bring a child into the world without a father. After discussion with the staff, Mr U was advised to change the form. He did this, initialling the amendments.

This meant that, should Mr U die or become incapacitated, the sperm would be destroyed. His wife would not be able to access it for treatment. Tragically, and unexpectedly, Mr U died after suffering an asthma attack. Distraught, Mrs U realized that she would no longer be able to pursue her course of treatment. She brought the case to court, claiming that the first consent form signed by Mr U should have been binding. She also suggested that he had been unduly influenced by staff at the clinic.

Despite sympathy for her situation, Mrs U's case was rejected not on the basis of the ethical concern, but on the grounds that her husband's final consent must hold sway.

Developments in reproductive technology have served to bring into question the relationships which have been regarded as essential aspects of parenthood and family life. They also raise concerns about the family structures that are necessary or desirable for children to thrive. Should we aim to ensure that children are born only into optimal family circumstances? If so, how do we go about identifying what *is* optimal? Clinicians are currently obliged to consider the need of a child for a father, but this is open to interpretation. Some clinics, such as the one in the case above, adopt an overt policy based on a moral belief. Others may come to a different view.

This may be difficult for prospective patients. But if establishing clinical harm is fraught with difficulty, establishing 'social' or 'family' harm is perhaps still more so. Questions relating to family structure are highly politicized and perceptions of what is desirable in this context are extremely subjective. Some argue that restrictions on family type are discriminatory, and should not be permitted. Others believe that it is utterly wrong to use technology to bring children into unconventional family environments, the effects of which are untested. Further discussion on family structure is included in greater depth in Chapter 7.

The non-identity problem

While avoiding harm seems intuitively desirable, interpretations of harm, whether clinical or social, can be subjective, and may lack clarity. It is not always obvious how to incorporate a desire to prevent harm into an ethical framework. And further to confound the issue, we have what philosophers have called the 'non-identity problem'.[21] Suppose a prospective parent is considering using an experimental fertility treatment such as reproductive cloning. The first reaction of observers may be to contemplate the possibility of harm arising to the cloned child. Indeed, the scientists tell us there is a high chance of physical defects, and possibly also of premature ageing, as well as other unforeseeable ill-effects.[22] Others suggest that the child would suffer psychological or emotional difficulties as a result of confusion over its family relationships.[23] In all, we might assume that to conceive such a child would certainly be to harm it, and therefore prohibit the use of reproductive cloning at least until it might be proved to be safe.

But this is too hasty: while a cloned child might well suffer physical or emotional problems, is it really correct to say that the child has been harmed? After all, if that child had not been cloned, then he or she would not exist. The choice is between existing as a clone and not existing at all. (Because, if the prospective parents decided to have a child naturally instead of cloning, a different child would be born.) It might be preferable to society or to the parents that a child should not be affected by any physical and other problems involved in cloning, but this cannot be attributed to a desire not to harm the cloned child, who would otherwise not exist.

This is not to say that it is better to exist as a clone than not to exist at all, but simply to point out that the problem cannot necessarily be couched in terms of harming the child. We cannot know whether it is better to exist than not to exist. Indeed if we think of existence as a benefit, this can result in equally problematic conclusions, since this could imply that the more people we bring into existence, the better.

Some people approach this dilemma by saying that the person has been harmed if, on balance, his or her life is not worth living. However, this leads directly back to the problem of quantifying suffering which, as we have already seen, yields no obvious answers.

Viewed from this angle, it seems that avoidance of harm is not after all the simple answer to the ethical problems raised by fertility treatments that it may at first appear to be. Yet intuitively, most of us retain our conviction that it would clearly be wrong deliberately to conceive a child knowing that the child would suffer from a disease. Clinicians and parents have a choice as the sequential steps of fertility treatments unfold. Can it be right to choose the risk of handicap or suffering? It seems clear that parents should and normally do want the best for their children, but actually interpreting and applying this in practice is fraught with difficulties because of the choices presented by modern genetics and reproductive technologies.

The closer one examines our moral assumptions, the more complex they seem. This leaves us in a difficult position when trying to decide where to draw appropriate boundaries. If philosophical certainty is impossible, total permissiveness also seems undesirable. Yet if boundaries *are* drawn, they must not appear arbitrary. One of the solutions to this problem is to try to ensure that the HFEA's moral stance within the law is at any rate consistent with public opinion and is flexible enough to adapt when that opinion changes.

When regulatory decisions are being made, there is rarely the opportunity for deep analysis. The HFEA operates within narrower parameters set by the law, and informed by pragmatic considerations. The role of the HFEA is not to dictate to society what its morals should be. It must take heed of public concerns, and ensure that, as far as possible, it reflects these social values. The key word here is 'social'. Some of the HFEA's hardest decisions have arisen from conflicts between the ethical beliefs of one individual and the interest of society in general. As individuals we may see no ethical problem standing between ourselves and the pursuit of our goals, but we are not always inclined to consider the outcomes of those goals in broader social terms. The HFEA has to try to achieve this duality; it is not equipped to find answers to the moral questions which have exercised philosophers for many years.

CHAPTER 3

Saviour Siblings, Designer Babies, and Sex Selection

Preimplantation Genetic Diagnosis (PGD)

The ability to control conception in the laboratory was welcomed by many women who were thus able to become mothers. But it also created new possibilities and applications. Scientists who were offering IVF usually obtained a number of eggs from women undergoing treatment in order to maximize the chances of success. Initially, 'selection' of which embryos to implant involved simply choosing those which were dividing normally. Any which were not developing, or looked 'odd' or asymmetrical, were discarded. However, it soon became clear that the existence of multiple embryos in the laboratory setting offered scope for further testing. Rather than simply observing the rate of cell division and symmetry of the embryos, scientists could go one step further and look inside the embryo's cells to ascertain the presence or absence of harmful genes.

Case 9: Susan and Chris Paget Dunthorne

Susan and Chris Paget Dunthorne suffered the loss of their first child when he was only a few months old.[1] He had been diagnosed with cystic fibrosis (CF), a serious genetic disease that attacks the lungs and digestive system.

The couple had not realized that they were carriers of this gene. Now, despite longing for another child, they feared that they would again pass on the disease. There was a test which could be carried out during pregnancy, but they did not want to go through the ordeal of aborting an affected foetus.

It seemed that Susan and Chris had no further options. Then they discovered that there was another possibility: they could undergo IVF and have the embryos tested before they were placed in the uterus. The couple

embarked on the process of ovarian stimulation, fertilization and genetic testing. Eventually, after a first failed attempt, Susan became pregnant. She gave birth to a healthy boy.

Families known to be at risk of certain genetic conditions were the first patients to be offered the chance of using the new technique. PGD can determine which embryos are affected by a particular gene before they ever reach a woman's womb. Embryo(s) that test clear of the inheritable disease can be transferred to the uterus, while those that carry the harmful gene are allowed to perish, or donated to research. In this way, families are spared the anguish of knowing that they risk giving birth to a child who may be devastatingly affected by a genetic condition. However, the negative association of the term 'selection', with its resonance of concentration camp and Nazi ideology, has awakened fears over the use of these techniques. Some also feel that the deliberate creation of embryos in the knowledge that some will be destroyed is questionable.

At present, the majority of couples seeking PGD have either had a child with a genetic disease, or a close relative affected by such a condition. Diseases which may be tested for include Huntington's disease, cystic fibrosis, chromosomal disorders, predisposition to certain cancers, and sex-linked diseases. The conditions tested for must be serious enough to merit the use of the technique. However, much of what is currently permitted relies on the state of genetic science. We are able to offer PGD only for those conditions whose genetic underpinnings are understood. There are many conditions, both serious and trivial, for which we simply do not have this knowledge.

This means that, rather than objectively deciding which conditions are good candidates for PGD and then seeking scientific means of establishing the genetic components, we have to wait to see which conditions *could* be tackled in this way. If genes for relatively trivial conditions were discovered (eg obesity or short-sightedness), it seems possible that some couples would want to use PGD to screen against them.

This raises one of the fundamental questions related to PGD: who decides which conditions are 'serious'? The statutory powers of the HFEA require that the distinction between serious and non-serious conditions is drawn in the first place by a public body rather than being dictated by the individuals concerned. Nevertheless, the HFEA Code of Practice relating to the factors to be considered in deciding whether PGD should be carried out gives back to the parents much of the determination of the gravity of the condition that is to be tested. Paragraph 12.3.2 states:

> The use of PGD should be considered only where there is a significant risk of a serious genetic condition being present in the embryo. The perception of the level of risk by those seeking treatment is an important factor in the decision making process. The seriousness of the condition should be a matter for discussion between the people seeking treatment and the clinical team.

It is interesting to note here that the distinction between serious and non-serious is not an objective one. It will fluctuate according to social needs and expectations, and according to the progress of science and medicine. For example, a condition such as type one diabetes would in the past invariably have been fatal since it was incompletely understood, and a means of obtaining or synthesizing insulin had not been developed. However, our current state of medical knowledge allows us to provide diabetics with treatments that enable them to live full and active lives. Thus, while once diabetes might have been seen as—literally—deadly serious, it is now eminently treatable. It is questionable whether, if PGD were available for diabetes, it would be seen as appropriate.

Another consideration here is the ways in which social expectations, and changes in lifestyle affect what might, or might not, be considered serious. Perhaps in our distant past, being a slow runner, or having bad eyesight, would have been mortally serious conditions, since we would have been unable to outrun predators, or spot lurking dangers. But in our current sedentary world, the need to outrun dangers has all but vanished, while contact lenses and glasses make bad eyesight an irrelevance.

Initially, the concept of serious disease was assumed to apply to conditions which would inevitably affect sufferers from early on in life. However, there are a number of genetic conditions which may not appear until adulthood or middle age, and some of these, such as Huntington's disease, are included on the list of conditions for which PGD can be used. The devastating effects of Huntington's disease are well-known. A process of slow mental and physical deterioration is inevitable, leading to eventual death. And because its penetrance (ie the likelihood of the gene actually causing the disease) is 100%, there are few to argue against the use of PGD to avoid conceiving children who will be affected.

However, at the other end of the spectrum are conditions for which the particular genetic test may reveal only a *predisposition* as opposed to a certainty that the carrier will suffer from the disease (also known as low penetrance).

Case 10: Chad Kingsbury

In September 2006, an article appeared in the *New York Times* under the heading 'Couples Cull Embryos to Halt Heritage of Cancer'.[2] The story described Chad Kingsbury, a man who had lost four family members to a virulent form of inheritable colon cancer.

Mr Kingsbury and his wife could not bear the idea that if they had children, they too would be at risk of the disease. Unlike some genetic conditions, however, the presence of the gene in question would not mean that the child would inevitably go on to develop the problem. Only a proportion of those carrying the gene would actually be affected.

This meant that the use of PGD might involve discarding embryos that would have led healthy lives—a hard choice to make. Nevertheless, for the Kingsburys, the knowledge that their child would be free from disease outweighed the other issues.

Through the use of PGD, they had a healthy daughter.

In 2004 inheritable colon cancer of the type described in this case was added to the list of conditions for which PGD could be licensed in the UK. Then, in 2006, after a lengthy review and public consultation, the HFEA decided that in principle PGD could be used for late-onset and low-penetrance conditions, even where such a condition may be treatable. So, for example, a woman whose family has an inherited susceptibility to breast cancer may be eligible to use PGD to ensure that her offspring do not carry the gene.

PGD offers hope to many, and the broadening of scope for its use means that greater numbers of people are likely to come forward in the future. But unfortunately, this hope is not always afforded without problems. PGD cannot guarantee that a child will be exactly as the parents wish. Diagnosis may not always be simple, and clinicians may make errors. Misdiagnosis may occur as a result either of the techniques involved or because the material removed for biopsy does not adequately reflect the makeup of the embryo.[3]

It is also important to remember that even if PGD screening for a particular cancer is successful, the child may nevertheless go on to suffer from the disease. This is because only a relatively small proportion of cancers are actually caused by inherited genes. Individuals conceived using PGD will share the same risk as other people of contracting cancer as a result of random mutations, or environmental factors. PGD, even when successful, does not entail that a child will enjoy perfect health. Parents who feel secure in the knowledge that they have 'insured' their child against the family disease may still find that the child suffers from a completely different, but equally devastating, condition. Even with the use of PGD, reproduction is always likely to be an uncertain endeavour.

Finally, since PGD requires the use of IVF techniques, there will be a relatively low rate of pregnancies for each cycle. It incorporates all the physical, emotional and financial problems of IVF and the very small number of viable offspring which result. Rather than guaranteeing a healthy baby, then, it might be more accurate to say that it merely makes it less likely that the prospective parents will have an unhealthy child, since the chances of a successful pregnancy may be significantly reduced.

The probability of getting pregnant from a PGD cycle is even lower than for a 'normal' IVF cycle. This is partly because PGD necessarily involves a reduction in the number of embryos available for implantation, as some of the embryos will carry the disease being tested for. Also, PGD itself may pose a risk to healthy embryos which have cells removed for biopsy, as this procedure can occasionally damage the embryo. When this happens the damaged embryo is discarded and this means that fewer are available for implantation.[4]

PGD raises some non-medical concerns too. One fear is that the technique may change social attitudes toward those who are born with genetic diseases. Parents who have failed to screen their embryos may be regarded as culpable, while a degree of opprobrium might also fall on the affected children. If these attitudes became prevalent, general willingness to support those born with genetic diseases might gradually be eroded.

Some people have also suggested that societal solidarity is threatened by the use of PGD to screen out embryos suffering from certain conditions.[5] So for example, if a couple decided to use PGD to avoid having a blind child, this could be felt by blind individuals to reflect a negative attitude towards the blind. It is argued that PGD in these circumstances expresses a view that those individuals affected by particular conditions should not have been allowed to be born, or that their lives are less valuable than those of other individuals. A related argument is that if fewer people with certain disabilities are born, society has a reduced imperative to find cures for these conditions.

Aside from this there may be other social concerns. One fear is that where there are fewer people who suffer from certain conditions, their voices are less likely to be heard, and they will have less political power as a group. It has been argued that disabilities are social rather than medical constructs: society fails to provide the mechanisms and facilities for some people to function fully.[6] A proponent of this view might regard the use of PGD to screen out disabilities as being pernicious. Instead of providing better services and remedying injustices, society would be seen to be trying to remove the people whose needs it should be meeting.

One counter argument to this is that where one is in a position to make a choice, it is surely natural to prefer that one's children are free of serious diseases or disabilities. But does exercising this choice imply any particular attitude towards people who *are* born with a particular condition? Making a decision as to what kind of child to have is a different proposition entirely from suggesting that certain people should not be alive, or that their lives are worth less than those of other people. Interestingly, while the use of PGD has been refined over the last decade, there has at the same time been more, not less, legislation which aims to give protection to disabled persons and, it seems, more compassion and understanding of debilitating conditions. Nevertheless, some people remain concerned at the idea that PGD represents an attempt to prevent certain people being born, rather than an attempt to find cures.

Positive selection

One of the greatest challenges involved in the use of PGD is the possibility that it could be used not to screen out genetic diseases, but to select for certain characteristics, such as hair or eye colour. Parents might want to ensure that their children would be tall and blonde, or that they would be of one or the other sex. Looking towards the mental characteristics, some parents might wish to look for embryos with particular intellectual qualities such as high intelligence or perhaps musical ability.

In all, it seems that PGD opens up many questions as to what kind of people we think technology should be used to create, what motivations parents should have when conceiving children, and how developments in these areas in which certainty is so hard to come by should be regulated. Any efforts to create logical boundaries between serious and non-serious conditions may be vulnerable to attack. Even distinctions between therapeutic interventions and enhancements are difficult to draw. Some people see no problem in permitting the use of technology to increase choice even when this extends beyond remedying diseases, while many others believe that guidelines and restrictions should be in place.

A number of public consultations have been held in order to ensure that HFEA policy on the use of PGD is in accordance with social concerns. Overwhelmingly, responses to these consultations have indicated that while the use of PGD technology to screen out disease is acceptable to the public, the idea of using it to create so-called 'designer' babies is not. The use of these techniques for trivial or non-medical conditions is therefore unlikely to become widespread in the UK, at least in the near future. However, the difficulty of establishing what constitutes trivial use remains a permanent feature of the ethical landscape in this area.

But there is still another question to consider in this context. What happens if parents want to use PGD to select embryos which do have a particular genetic disorder? It might be assumed that such a bizarre wish would never occur—why would parents want to harm their offspring? However, there have been cases where prospective parents have sought to conceive a child with what would widely be regarded as a disability.

Case 11: Sharon Duchesneau and Candy McCullough

A Canadian couple, Sharon Duchesneau and Candy McCullough, both deaf, wanted to have a child.[7] The couple needed to use sperm donation in order to conceive, and decided to maximize their chances of having a deaf child by asking the sperm bank to provide sperm from a deaf donor.

However, the couple were informed that the donor bank operated a screening process meaning that if any donors *were* genetically predisposed to deafness, they would be automatically turned away.

On learning this, the women took matters into their own hands and made contact with a potential donor themselves. Their chosen candidate was deaf, and came from a family affected by deafness over several generations.

The pregnancy went according to plan, and Ms Duchesneau gave birth to a baby girl. The child, as hoped, was deaf. Some years later, Ms Duchesneau and Ms McCullough felt that they wanted to complete their family, and again sought to conceive a child. On this occasion, their desire to conceive a deaf child was picked up by the media, provoking a hostile backlash from many commentators who were shocked at what they saw as deliberately designing a baby with a disability.

However, the couple went ahead with their plans, and their son was born as a result of insemination from the same donor who had provided sperm for their daughter. Again, as planned, the child was born with severely impaired hearing.

The reasoning behind this couple's desire for a deaf child was based largely on the fact that there is a lively deaf community to which they thought it was desirable that their offspring should belong. These parents did not think that, in selecting deafness, they would be harming their child, but that they would be providing that child with the necessary attributes to participate in their community. Because they were both women, they were obliged to seek reproductive assistance in order to have a family at all, but it is interesting to consider that, had this been a fertile, heterosexual, deaf couple, they would have been free to conceive—naturally—a child who they hoped and expected would be deaf. Perhaps part of what they were aiming to achieve was the conception of a child which would be as near as possible to the child they might have had together.

Similar scenarios might be imagined with reference to achondroplasia (dwarfism), where parents might feel that they wished their offspring to share their identity and community or, on a more prosaic level, might simply be aware that their modified living environment would be ill-suited to a child of normal stature.[8]

While the case described above did not involve the use of PGD, it demonstrates the huge variation in what people think is reasonable in terms of desiring specific attributes for one's offspring. In the current climate, then, PGD is not simply a tool for remedying disease; rather, it is an ethically-fraught instrument which may serve to test social values and assumptions and even our understanding of what is a disease and what is not.

Screening for carriers

The use of PGD has much scope for generating uncertainties and controversies. This is why it is allowed only under strict controls. This includes insisting that clinics make arrangements for genetic counselling and give information to the patients with an explanation of the risks involved. The decision to use PGD will be made in consideration of the unique circumstances of those seeking treatment, not simply on the basis that they are carriers of a particular genetic condition.

Some genetic diseases, such as Huntington's, are dominant. This means that anyone who has a copy of the gene will suffer from the condition. However, other conditions, such as cystic fibrosis are recessive. This means that people with only one copy of the faulty gene will not suffer from the illness, and may be unaware that they are 'carriers' of the gene in question. When one carrier reproduces with another carrier, there is a one in four chance that the child will suffer from cystic fibrosis as a result of having inherited the faulty gene from both parents.

Where couples know that they are carriers of the cystic fibrosis gene, they may decide to use PGD to select an embryo free from cystic fibrosis. If they take this route, there may also be a choice as to whether to choose an embryo entirely free from the cystic fibrosis gene, or one which carries the single faulty copy of the gene. Given this choice, many parents would opt for the embryo which is completely free of the gene.

In some cases, however, this may lead to unforeseen problems. Sometimes genetic mutations can carry certain advantages. If we eradicate all the carriers of that gene, we may simultaneously lose the associated advantage. An example of this can be seen in the case of sickle cell anaemia. Like CF, sickle cell anaemia is a recessive disease which only affects people if they carry two affected genes (ie one from each parent). But whereas there is no known benefit from carrying the single faulty gene for CF, people who carry only one copy of the sickle cell anaemia gene often have a degree of immunity to malaria.

Because we know that being a sickle-cell carrier has this benefit, parents and clinicians can consider carefully before using PGD to screen against carriers. However, we do not know how many other genetic mutations may also convey benefits, which would be lost if we screened out carriers. This might argue for caution where screening for carriers is in question.

Sex selection

Case 12: Alan and Louise Masterton

In 2001, Alan and Louise Masterton[9] travelled to Rome and paid £6000 to an Italian fertility clinic which had agreed to help them select the sex of their next child. Alan and Louise already had four sons. They had lost their only daughter in a tragic accident in 1999, and were desperate to have another little girl.

The couple were adamant that they were not trying to replace their daughter, but simply trying to re-establish a female element in their family which had been lost with their daughter's death.

The HFEA refused to allow the couple to have the sex selection procedure performed by British clinics, although sex selection is permitted if there is a known risk of the parents transmitting a sex-linked genetic disease. To allow sex selection for social, as opposed to medical reasons, was regarded as unacceptable.

The Mastertons argued that this simply forces people to go abroad and seek treatment—effectively limiting this option to those who can afford it. As it transpired, the Mastertons' attempt at conceiving a girl with the help of the Italian fertility clinic failed. They produced one viable embryo, but it was male. They donated it to another couple.

In 1993 a clinic opened in London offering sex selection by sperm sorting. The HFEA was concerned, and decided to hold a public consultation on this issue. The question was: should sex selection be allowed only for medical reasons, or should it also be available for social reasons? Underlying this is the question of whether sex selection for social reasons is intrinsically wrong. The clinic which advertised sex selection services in 1993 was beyond the HFEA's regulatory control, since it was not using IVF, nor was it storing sperm or eggs. (Arguably this may be seen as a flaw in the regulatory framework. Perhaps sex selection *should* be a matter for the regulator regardless of the methods used.) However, even though this particular clinic was not answerable to the HFEA, it was clear that where IVF and PGD are used, the option of sex selection is at least technically feasible. It became important that the HFEA should develop some kind of stance in relation to this possibility.

Accordingly, a public consultation was launched, but surprisingly few people responded. 2000 copies of the consultation were sent out, and only 165 responses were received. Nevertheless, the results were informative, with 67% of respondents being opposed to the use of sex selection for social reasons. The reasons given included the idea that sex selection could reinforce gender stereotypes, and fears that it might lead to a 'slippery slope' whereby further attributes might be selected, with children being viewed as 'consumer goods'. Detrimental effects on family life were also feared in the event that the technique was unsuccessful. Other concerns raised included questions of whether social sex selection was an appropriate use of medical resources and technology.[10]

In keeping with the findings of the consultation, the HFEA concluded that sex selection for non-medical reasons was inappropriate. This has been criticized by some commentators who feel that restricting the freedoms of adults on the basis of such a small sample of views is unjustified.[11] Where responses are invited on an issue, it is likely to be those with strong views who are motivated to contribute, and arguably, this may skew results. However, the fact remains that if the rest of the public have not been motivated to protest, or to register their views in favour of sex selection, there seems little point in overruling the majority view of those who *did* respond.

But how compelling are the arguments raised against sex selection? In the past many women who hoped to have a boy have given birth to a girl. This might have been distressing, annoying, or devastating, depending on the nature and motivations of the parents, but they could not *blame* any particular individual for this. If sex selection is allowed, human beings would be given responsibility for something which has hitherto been random. This might not be problematic if the technology were 100% effective; however, this is not the case.

If parents choose sex selection, and find that it has failed, there might be a tendency to blame the clinicians who have failed in their promises. There might also be a temptation to reject the child, or—if the error is found out prenatally—to abort the foetus. The effects of sex selection on family relationships are difficult to foresee. Perhaps parents who had already had one child of the 'wrong' sex and then succeeded in using sex selection to have a child of the 'right' sex would favour the latter. Children of the unwanted sex might grow up feeling bitter and resentful.

Of course, many of these problems might occur even without sex selection, where parents have a strong desire for a child of one or the other sex, and their offspring does not fulfil that desire. Perhaps the underlying reason for many people's opposition to sex selection is the idea that if parents specify a desire for a particular sex, this indicates a far from unconditional approach to parental love. In reality, unconditional parental love may be an unrealistic ideal. Whether or not they use reproductive technology to meet the desire for particular attributes in their children, many parents have such desires. However, this does not necessarily entail that such preferences should be sanctioned or facilitated.

The initial consultation on sex selection took place in 1993. In view of the development of more sophisticated techniques for sex selection and possible changes in public opinion, a subsequent public consultation was held in 2002/03. Despite the passage of time, it appeared that public opinion had not changed significantly. According to the HFEA report, a MORI poll conducted as part of the consultation revealed that 69% of people thought that sex selection for non-medical reasons should not be available.[12]

Many other countries around the world take a different view. In Israel sex selection is permitted by a committee on a case-by-case basis.[13] One anecdotal case in which permission for the use of sex selection was granted involved an Arab couple. After having had eight daughters in succession the couple had become objects of scorn in their village. It was feared that these daughters in turn might be regarded as unmarriageable because of a widespread belief that they would produce only female children.

On a global level, it has been suggested that the use of sex-selection might disrupt the natural proportions of male-female ratios with unforeseeable consequences. In Britain and the US, studies have indicated that there would not be a marked preference for boys over girls—one of the fears that was initially associated with sex selection. However, it is indisputable that some cultures do show a marked preference for male children. Providing further opportunities for people to select sex, even if we think that we ourselves are free of those injustices (in itself a questionable claim) might further the gender injustices that exist around the world.

The problem with cases such as the Mastertons is that on an individual basis they are compelling, but the regulator is obliged to look beyond the individual to the broader social and ethical implications of the choices being made. As the Mastertons themselves pointed out, this may result in people who can afford it seeking treatment elsewhere. Again, this situation is not ideal, but the alternative may not be any more desirable.

If we permitted all the treatments and procedures which are allowed in at least one other country in the world, it would be unlikely that we could maintain laws against, for example, reproductive cloning or the creation of animal-human hybrids. If any ethical credibility and independence is to be maintained, there must be times when national convictions that a practice is wrong must prevail. Cultural and ideological contexts will differ from country to country, and even if we accept that, for example, sex selection should be permissible in Israel, this does not inevitably mean that we are obliged to legalize it in the UK.

The use of sex selection is permitted for medical reasons in the UK. Some families carry genetic mutations which can cause serious and debilitating diseases for a male child, but which would not affect a female child. Where this is the case, parents may wish to ensure that they do not run the risk of having an affected child, and this can be achieved by selecting only embryos of a particular sex for implantation. This seems to present a very different scenario from that involved in so-called social sex selection. However, boundaries between the social and the medical are not always entirely clear-cut.

It is perhaps inevitable that where boundaries have to be set up, there will be a degree of uncertainty as to whether the division between what is permitted and what is not, is anything more than arbitrary. This may be a good argument for revisiting such boundaries, and going back to the public to ensure that regulations do not contradict society's views on issues such as these. Public views do change, and for this reason, regulation needs to be flexible.

Saviour siblings

One of the most controversial uses of PGD has been to enable parents to select an embryo which, if carried to term, would result in a healthy baby whose tissue or umbilical cord stem cells could be used to cure an older sibling affected by disease.

Case 13: The Hashmis[14]

Raj and Shahana Hashmi, a couple from Leeds, became a media sensation after it was revealed that they were seeking to use PGD to create a 'saviour sibling' for their son, Zain, who suffered from a blood disease called beta thalassaemia. This affects sufferers' ability to produce haemoglobin and red blood cells, and causes such severe physical symptoms that, without lifelong medical care, life expectancy would be only a few years. Zain's quality of life had been blighted by the painful nature of the required treatment (repeated drug infusions and blood transfusions).

Ironically, Mrs Hashmi had been aware that she was a carrier of the genetic mutation, and had therefore undergone prenatal testing to ensure that Zain would not be a sufferer. Tragically, on this occasion the test failed, and Zain was born with the disease.

By the time Zain was 2½ years old his symptoms had become extremely severe; his treatments were highly invasive, and his life expectancy was uncertain. His parents were desperate to spare their son this suffering. They hoped that if they conceived another child with the right tissue-type, cells from that child's cord blood (blood taken from the umbilical cord at birth) would achieve a cure, relieving Zain from the lifelong cycle of drugs and transfusions. None of Zain's three existing siblings was a suitable tissue match for him.

Initially, the Hashmis tried to conceive naturally to provide a tissue-matching sibling for Zain. However, on the first occasion, prenatal testing showed that the embryo would also have had beta thalassaemia, so the pregnancy was terminated. On the second occasion, Mrs Hashmi conceived a healthy child and gave birth to a son. Unfortunately, the child was not a tissue match for Zain.

The Hashmis then sought permission from the HFEA to use PGD to ensure that their next child would both be free of beta thalassaemia, and would be a tissue match for Zain. The HFEA believed that in this family's particular circumstances, the creation of a 'saviour sibling' was indeed justified. However, there was a degree of public unease in respect of this, and this was heightened by a court challenge to the HFEA by the pro-life organization 'Comment on Reproductive Ethics' (CORE). The challenge was based on a claim that the HFEA was not empowered to issue PGD licences for the use of tissue typing to select between healthy embryos. The HFEA lost the case at the first instance. However, on appeal, this judgment was overturned, and it was found by the Court of Appeal and the House of Lords that this kind of tissue typing *was* something which the HFEA was entitled to license.[15]

Sadly the Hashmis' attempts to conceive a tissue-matching sibling for Zain ended in a succession of miscarriages. With the passing of time, the chances of conception grow ever smaller and eventually, repeated attempts at PGD and IVF were thought not to offer a high enough chance of success. The Hashmis launched a nationwide appeal for possible matching donors to come forward for Zain, and others like him. They continue to press for increased donation, especially in the Asian community.

The case of the Hashmis demonstrates the ethical and scientific complexities of PGD. As previously stressed, IVF is in itself a precarious endeavour, and any one cycle carries a relatively small chance of pregnancy. For the Hashmis, therefore, the success in court could not guarantee that they would in fact succeed in creating a 'saviour sibling' for Zain: it merely allowed them to try.

So how did the HFEA approach the complex arguments surrounding the Hashmis' case? It started from the premise that that which was not forbidden in general should be permitted (although there are those who say that the very existence of the HFEA is a denial of that principle). This was followed by an examination of the case in the context of the usual ethical grounds applied by the HFEA. Here the emergence of a possible new principle was noteworthy: the saving of life. The request was considered carefully over a number of months, and close attention was paid to the wording of the original HFE Act. Section 13(5) of the HFE Act stipulates that:

> a woman shall not be provided with treatment services unless account has been taken of the welfare of any child who may be born as a result of the treatment (including the need of that child for a father), and of any other child who may be affected by the birth.

In the edition of the HFEA Code of Practice that was current at the time, clinicians were advised that:

> When considering the treatment of any woman, treatment centres must take into account the welfare of the child that may be born as a result of treatment. Treatment centres are expected to also consider the welfare of any children the woman may already have responsibility for and the effect that treatment could have on these children. Treatment centre staff are expected to be aware of the need to show both care and sensitivity in this decision making process. Consideration is expected to be taken regarding the wishes and needs of those seeking treatment and the needs of any children involved.[16]

The wishes and needs of the people seeking treatment in this case (ie the Hashmis) were abundantly clear: they wanted a child who would be able to help their son. In terms of the effects of treatment on existing children, this too seemed relatively straightforward: Zain would benefit immeasurably from the birth of a tissue-matching sibling.

The HFEA also considered Article 8 of the Human Rights Act 1998 (the right to respect for private and family life), Article 12 (the right to marry and found a family), and Article 14 (the prohibition of discrimination). There is a presumption in law that people should be free to exercise their rights in areas of activity that most closely affect themselves and their families. It is accepted that a public authority should only interfere with the exercise of these rights in the interests of public health or morals. Where it does interfere, this interference must be proportionate and the presumption for intervening should be strong.

Given that the choice to select an embryo with certain characteristics was a technological possibility, it could be argued that to prevent the Hashmis' use of it would have constituted a restriction of their freedom—or an infringement of their right to family life, for example. This might seem to speak in favour of allowing them freedom to go ahead. However, it became clear that if the question was addressed purely from this angle, the rights of parents in general could be argued to outweigh any restrictions on access to reproductive technology.

If parents' privacy can justify their selection of their children's attributes, then would selection of blonde hair or deafness also be justified? Perhaps not. Parents' rights in this respect are not absolute: they can be constrained by other relevant considerations. The question was now to establish whether any such relevant considerations applied here. The most important concern was the issue of whether PGD with HLA tissue typing (the match required in the Hashmi case) was compatible with the welfare of the unborn child. Would a child conceived as a tissue-match for Zain suffer, or experience ill-treatment as a result of having been 'selected' in this way?

There were several possible ways of looking at this question. One was to weigh up the physical dangers involved in the processes of PGD and HLA tissue typing itself. These were found to be no more than any other use of PGD, as although PGD may endanger the embryo at the time of testing, any damaged embryos are not implanted, so the child that is born will not suffer from this damage.

Thus, the likelihood of a child conceived in this way being physically damaged or diseased due to the process of the technique was not significant. But would the process of cord blood donation harm the child? The easy answer to this question is no: the cells involved are not removed from the child, but from the umbilical cord after it is cut, so this has no direct impact on the child. Where cord blood is not collected after a birth it is simply discarded. There was, however, in some people's minds another form of physical danger which might affect a child born in these circumstances: namely that further—harmful—demands might be made for the child to undergo medical procedures for the benefit of the older brother.

Suppose, for example, Zain suffered kidney failure a few years after the birth of his new sibling. The existence of a perfect tissue match within his own family might result in the assumption that the younger child should donate a kidney to his brother. However, any claim on a child's bone marrow, kidneys or other organs would have to be approved by a judge. In any case where surgical intervention is proposed which is not directly for a young child's own clinical benefit, a court ruling must be sought. The rights of a saviour sibling with respect to any future treatment would be identical to those of any normally conceived child.

But there is a further consideration to address here. It is often stated that people should be treated as ends in themselves, never merely as means to the ends of others. It was here that much of the unease over the prospect of 'saviour siblings' seemed to lie. Unarguably, a child conceived in order to provide tissue for a sick sibling is conceived as a means to an end: the end of curing the sibling. For many people, this seemed to be the end of the matter: the proposition was an unacceptable violation of the rule, and must not therefore be allowed.

However, this is an oversimplification. While it might seem possible to accept that we must never treat people merely as a means to an end, we cannot avoid often treating people partially as means to an end. If Jane invites John for dinner, John might be using Jane as a means of obtaining a meal. To this extent, he uses her as a means to an end. But if John also regards Jane as a human being with her own motives for her actions, he is not treating her solely as a means to an end by accepting her invitation. If John arrived unwanted, uninvited and forced Jane to cook a meal for him, and afterwards left without having made conversation or offering to help with the washing up, perhaps this would seem more clearly to indicate that she was being treated as a mere means.

The complexity of untangling means and ends is still more difficult when it comes to conceiving children. It may be impossible to discover parents' true motivation for wishing to have a child. Often, they may not fully know themselves why they are driven to have children. In fact, to enquire into people's motivation for having children is a very modern preoccupation. From the beginnings of human history until the end of the twentieth century, control over how, whether or why to have children was limited. Even in these days of choice, it is not necessarily incompatible with the welfare of the child that the parents' desire for a child for the child's own sake is not their primary or sole motivation.

People may conceive a child in order to provide a companion for an existing child, or to give their lives a sense of worth, or to give a grandchild to their own parents. In countries where childbearing is actively encouraged, people may reproduce in order to be eligible for tax concessions. Holders of hereditary titles may reproduce in order to have an heir. Some of these motivations may seem more or less worthy than others. What is clear is that to a degree all children are conceived (unless simply by accident) partially as a means to an end. When a parent decides to have a child, it cannot be said that the child is an end in itself, since that child does not yet exist. Rather, the parent seeks to have the child as a means of fulfilling a desire (an end) of his or her own.

The measure of good parents is that they love their child regardless of whether or not that child meets the ends which they had in mind when deciding to reproduce. If the Hashmis had loved and nurtured a new child and regarded him as a worthy person in his own right, they would not have been treating him merely as a means even though he had been conceived in order to meet their own ends. There could be parents who would not do so, and who would indeed treat a saviour sibling as a mere means. But this would depend on the character of the individuals involved: it is not a necessary corollary of conceiving one child to save another.

Questions still remained as to whether allowing the procedure was compatible with the public good and whether morally significant criteria could be found to demarcate acceptable and unacceptable reasons for the conception and selection of embryos. That is to say, there was a possibility that allowing this procedure could lead to the use of PGD in other contexts which were unacceptable.

For this reason, the HFEA drew up a list of considerations which must be addressed in licensing PGD in these circumstances.

- Firstly, the condition suffered by the existing child must be severe, life-threatening, or of sufficient seriousness to justify the use of PGD.
- It was also felt to be important that the embryos involved should also be at risk of the genetic condition which was suffered by the sibling

(as was the case with the Hashmis; all of their children were at risk of inheriting the disease).

- Other treatment options should have been fully pursued before turning to the 'saviour sibling' possibility.
- None of the PGD/HLA techniques should be available where the intended beneficiary is a parent rather than a sibling.
- Only cord blood should be taken—no other organs or tissues.

In laying down these guidelines, it was hoped that the risk of this technology being mis-applied, or being used for trivial purposes, was mitigated.

The procedure of PGD failed for the Hashmis, but their case nevertheless created a strong effect on the rest of society. The term 'saviour siblings' was coined, and the arguments about 'designer babies' resurfaced with more heat than ever. In the wake of the court ruling however, a significant precedent had been set. Other people with sick children suffering from inherited genetic conditions could now benefit from the ruling which had come about through the Hashmis' battle, and it was hoped that in subsequent cases the procedure might result in a positive match.

For the HFEA, meanwhile, a case still more perplexing than that of the Hashmis was about to arise; a case which would challenge many of the principles on which the Hashmi decision had been based.

Case 14: The Whitakers

Michelle and Jayson Whitaker, a couple from Derbyshire, approached the HFEA with a request to conceive a 'saviour sibling' for their son, Charlie.[17] Charlie suffered from Diamond Blackfan anaemia, a very rare condition which causes a deficit of red blood cells. Fewer than 1000 individuals are known to suffer from the condition worldwide, and the cause is not fully understood. Treatment is arduous and invasive, involving repeated transfusions. In Charlie's case, the treatments were having a severe effect on his wellbeing. A bone marrow transplant from a matching donor was Charlie's best hope. However, Charlie's sister—the most obvious potential donor—was not a match for him.

Charlie's parents wanted to have another child to complete their family. They decided to try to ensure that the new child would be a good tissue match for Charlie, so that blood cells from the umbilical cord could be used to cure Charlie.

The Whitakers and their clinic put their case to the HFEA, which carefully considered the issues involved. After much deliberation, the HFEA concluded that the use of HLA tissue typing in this case was not justified. The clinic was refused a licence for the procedure, and the Whitakers were informed that they could not pursue this course of action. Devastated, the Whitakers expressed themselves unable to understand the HFEA's reasoning. They decided to go to Chicago to seek treatment there. This treatment was successful, and in 2003 Mrs Whitaker gave birth to Jamie, a perfect match for his brother.

The Whitakers could not see why the HFEA were willing to licence the use of tissue-typing in the case of the Hashmis, but not for them. They were inclined to think that there were ulterior political or pragmatic motives for the decision in their own case. This assumption was strengthened by the response of the media, which took the position that the HFEA were cruelly thwarting a child's only chance of a cure. This was in contrast to the disapproving stance of the media when permission was granted to the Hashmis.

The different aspects of the Whitaker and Hashmi cases were extremely challenging both for the HFEA and for those who tried to unravel the legal and moral complexities involved. Certainty in these issues is unremittingly elusive. What *is* certain is that the HFEA agonized over these deliberations, and there was nothing arbitrary in its decision, even if this decision may justly be questioned with hindsight. At the time, the judgement was made in good faith.

Because of the obvious similarities between the positions of the Whitakers and the Hashmis, it was not necessary to revisit the ground covered with reference to the welfare of the child, and the philosophical questions of whether such a child would be regarded solely as a means. However, there was one important difference between the two cases, and it was this difference on which the HFEA's decision hinged. While the Hashmis risked giving birth to a child with beta thalassaemia if they conceived without PGD, the Whitakers did *not* run such a risk. Charlie Whitaker's disease was caused by a random genetic mutation, and his siblings were no more likely to be affected by his condition than any other child in the population.

This meant that the Hashmis were already eligible to use PGD to avoid having a child with the same disorder as Zain. However, the Whitakers had no reason for using PGD apart from their aim of having a child who would be compatible with Charlie. It was believed that PGD imposes risks on the embryos involved. These risks may be acceptable where it is important to ensure that diseased embryos are not implanted. However, where the technique is used solely to select between healthy embryos, none of the embryos receives any therapeutic advantage, but all are subjected to the risk. This fact was at the heart of the HFEA's stipulation that tissue typing should only be used where the child to be born is itself at risk of the condition being screened for.

This distinction has been criticized as being based on false logic.

The HFEA's judgement was partially based on the idea that it was unacceptable to impose the risks of PGD on an embryo when the procedure would confer no therapeutic benefit to that embryo. The risks were being imposed purely for the benefit of someone else. In the Hashmi case, there was a therapeutic benefit to be gained by any embryo selected: namely, the benefit of being free from beta thalassaemia.

But in reality none of the embryos involved in PGD receives any therapeutic benefit. PGD does not cure genetic disease, it merely enables diseased embryos to be identified and discarded, leaving only unaffected embryos for implantation. Healthy embryos receive no therapeutic advantage from this procedure since they were never affected by the disease. The only benefit such an embryo may gain (if it can be termed a benefit) is to be selected from amongst the others for implantation. Clearly, any diseased embryos do not benefit from PGD either, since they are discarded.

This being the case, it is not clear that an embryo created for the Hashmis would have benefited more from being subject to PGD than an embryo created for the Whitakers.

If there is a distinction to be made between the two cases, it cannot be done on the grounds that an embryo is receiving a therapeutic benefit on the one hand, but not on the other. However, the HFEA's position was also based on a broader argument about the general purpose of PGD. If PGD is used primarily to prevent children being born with certain disorders, it seems entirely plausible that its use should only be considered in these cases. Where it is being used for such a purpose, a secondary purpose may be added: namely, tissue-typing. PGD was not a procedure to be used for tissue-typing or for the selection of desired traits alone. The primary goal of PGD was to prevent the birth of individuals suffering from serious genetic conditions.

The pressure of public opinion, and the complexities of the issues involved placed great pressure on the HFEA to modify its judgement. New HFEA guidelines no longer restrict PGD for the selection of a saviour sibling to cases of genetic disease.

The two cases discussed here demonstrate the difficulty of dealing with this tide of opinion and rhetoric which may often be highly emotive, and frequently self-contradictory. The HFEA's decision on the Hashmi case, for example, was decried in the media. 'Quango in designer baby row', read one *Daily Mail* headline.[18] Despite this negative coverage, the refusal in the Whitaker case was also condemned and this time the HFEA was reprimanded for not showing a more liberal and compassionate approach.

This is part of the perennial problem for the regulator. It seems inevitable that for any decision, a vocal cohort of dissenters will emerge, while those who might have agreed, or are indifferent, may remain silent. This means that media attention and public perceptions are nearly always polarized in a way which makes dialogue fraught. Despite this, it is essential that the HFEA does not merely pander to what appears in the press. Awareness of public feeling is vital, but over-hasty capitulation to media pressure is not the appropriate way of incorporating public opinion into policy.

One possible way to deal with adverse media coverage in the wake of cases such as the Hashmis and Whitakers would be to return to the HFE Act, in order to clarify in legal terms exactly what circumstances justify the use of PGD and embryo selection. This would mean that the scope for the HFEA and members of the public to interpret the law in different ways would be diminished. Arguably, this could simplify the situation both for the public, who would benefit from clear legal distinctions between what is and is not permitted, and for the HFEA, which would not be forced into making agonizing ethical judgements over decisions which do not fit neatly into the existing legislation.

However, one of the problems with this suggestion is that the law is a blunt instrument in the context of fast-moving research and volatile public opinion. Changing the law with every new development to allow for new cases is not necessarily a good idea. Detailed amendment also erodes the flexibility that has allowed the HFEA to function day-by-day over the past 15 years, coping with medical issues that need immediate answers. It is unlikely that any amendment to the HFE Act intended to govern the use of PGD would be as subtle and adaptable as the discretionary approach developed by the HFEA over a number of years, with the assistance of public consultation.

CHAPTER 4
Fertility is a Feminist Issue

There are many different concepts of feminism, not all of which are addressed here. The definition and values attached to the term 'feminism' have changed over the decades, or even shorter periods, in the minds of the public and of women themselves. This chapter's approach to feminism focuses on women's autonomy and assisted reproduction. In this context the hallmark of autonomy and the integrity of the self is the right to make decisions about one's own body and to regulate one's own fertility.

Artificial reproductive technologies have been regarded by some as an unquestionably beneficial development for women. It has even been suggested that new reproductive technologies such as surrogacy and cloning could fundamentally change the relationship between reproduction and the alleged oppression of women.[1] In the past, one of the most pressing problems faced by women in their fight for equality with men was their inability to control their fertility. This meant that they were slaves to their biological make-up. Today, we take for granted much of this control, which was lacking for previous generations of women.

This is not to say that contraception was unknown in the past.[2] Indeed there were many ingenious ways of attempting to prevent pregnancy, some more effective than others. Condoms of various materials have been in use from the very earliest days. But for condoms to be effective, a woman must rely on the co-operation of the man. And while some of the more arcane techniques known to women may have had a limited degree of efficiency, women were simply not able to control their fertility as they are today. Women's attitudes toward this situation varied, but by many it was regarded with rueful resignation. After all, there was little they could do about it. 'I know it must be very horrid to go, as it were, a beast to the shambles', wrote Josephine Butler to a woman about to be confined, in 1893.[3]

With the advent of the contraceptive pill this suddenly changed. The pill enabled women to avoid having babies until they were ready for motherhood. This seemed to open new possibilities for women to build successful careers around timely reproductive choices. Although this belief turned out to be simplistic, it gave a generation of women the confidence to forge ahead and choose their career paths as if they were as unencumbered as men. The development of reproductive technologies could be regarded as a further step along this path to equality in men's and women's reproductive endeavours. But this supposed equality has had some devastating consequences.

Equal interests in reproduction?

Parenthood can now begin in the laboratory. The necessary connection of the embryo with its genetic mother's body has been broken. A woman genetically unrelated to the child can gestate and give birth to the child. This seems to make the mother's role more like that of a man: merely the provision of a gamete. From this perspective, reproductive technologies may look like a boon in every way, promoting freedom and choice for women as well as men in reproductive terms. However the equal weight given to men and women in disputes over *in vitro* embryos may be regarded as far from beneficial to the women concerned.

Case 15: Natallie Evans[4]

Natallie Evans was devastated when she was told that she had cancer. Not only was she faced with the difficulty of battling the disease, but the treatment involved the removal of her ovaries. Ms Evans decided to have some of her eggs removed prior to the cancer treatment, and fertilized with her partner's sperm. The embryos were kept in storage while she underwent her treatment.

However, the relationship broke down, and Ms Evans' partner decided that he no longer wanted to have a family with her. He requested that the couple's embryos be destroyed. Ms Evans embarked on a lengthy court battle to save her embryos, and her right to implant them. At each successive turn, she was turned down, despite the sympathy that judges had with her case.

In April 2007, her final appeal was rejected.

This conflict between Ms Evans and her ex-partner was extremely difficult to resolve. The outcome was doubly bitter for Ms Evans since her former partner still has the capacity to have children with another woman, should he so wish, while her chances of having a genetically related child have vanished.

Had Ms Evans been able to conceive naturally, her partner would have had no claim to demand that she abort the embryo(s). But when embryos exist outside the woman's body it is no longer clear that the woman's preference as to how they are to be dealt with should have greater weight than the man's. For many the judgment was welcome as an indication that fatherhood is taken as seriously as motherhood and that reproductive technology is not to be allowed to reduce the role of men to that of mere fertilization. However for Ms Evans it was a personal tragedy.

A similar case has arisen in the American courts.[5]

Case 16: Junior and Mary Sue Davis

Junior Davis and Mary Sue Easterly met while serving in the US Army. They got married in 1980 and a few years later were ready to start a family. However, Mary Sue's conceptions all ended in failure. She suffered a succession of ectopic pregnancies (where the embryo implants outside the uterus), leaving her with permanent damage to her fallopian tubes.

Eventually, she sought fertility treatment. However, a number of attempts at IVF also failed. Finally, the couple decided to create and freeze embryos rather than having them implanted straight away. The embryos could then be thawed and implanted at an optimal time in Mary Sue's natural cycle.

However, in 1989, Junior Davis filed for divorce. The status of the embryos became unclear and both parties claimed the right to decide what should be done with them. Initially, Mary Sue wanted to continue with her plans for pregnancy with the embryos, but her husband opposed this. As time passed, the wishes of the couple changed, and Mary Sue sought to donate the embryos to another couple.

After lengthy court proceedings, the embryos were destroyed.

In Israel, a strongly pro-family culture, the decision in a dispute over embryos went the other way. In the case of *Nahmani v Nahmani*[6] it was found that the provider of the sperm which had fertilized the eggs must not obstruct a woman's chance of having children. Interestingly, the ruling was based on the premise that a woman's interest in being a mother outweighs a man's interest in not being a father. Mrs Nahmani was probably pleased with this outcome, but whether it should be welcome to feminists is another matter.

Should we regard women as having a greater interest in reproduction than men? Or should we seek equality, upholding the interests of both adults who make a genetic contribution to their offspring? Clearly, when reproduction involves the mother's body, there is an additional consideration since any attempt to intervene in her choices would involve physical violation. But when embryos have been created and are being stored *in vitro*, a woman can no longer claim special rights over them based on the fact that they are in her womb.

To suggest that, solely by virtue of their sex, women have greater interests in becoming parents than men do, could set a reactionary train of events in motion. After all, if we agree this, then why not also agree that careers are likely to be less rewarding for women? On this view, men, free of the parental longings that afflict women, might well be expected to throw themselves more wholeheartedly into their careers. Extrapolating from these differences, it might seem sensible to return to the idea of enforced separate roles for men and women: the former in the workplace and the latter cosily ensconced at home with the children.

No doubt there are some who would approve such a move. However, even if it is admitted for argument's sake that most women place a high importance on having children, it is not clear what action can be justified on this ground alone. As long as there are *some* women for whom this is not true, it makes little sense to constrain an individual's options solely on the grounds of gender. Rather, judgements should be made on that particular individual's aptitude or desire for parenthood.

Likewise, if men are generally less driven to become parents than women, there may nevertheless be some for whom the desire is equal if not greater. Thus, while in the individual case of *Nahmani v Nahmani*, it might well have been true that in that particular case the woman's desire to be a mother was greater than her husband's desire not to be a father, it would seem highly dubious to infer any overarching principle from this. Certainly in the UK, any temptation to regard the woman's right to motherhood as paramount has been avoided in the law.

The case of Ms Evans shows that even where there is the technological ability to remedy infertility, social and legal concerns may nevertheless override a woman's desire to become a mother. This is an important consideration to bear in mind: technology alone cannot solve problems which are bound up with social and legal proscriptions. It is not always possible to foresee the social and legal complexities that may arise from apparently benign and uncontroversial techniques.

Egg freezing

Case 17: Helen Perry

In 2002, the first 'frozen egg baby' in the UK was born.[7] Helen Perry and her husband Lee had sought fertility treatment after several years of marriage. It had transpired that Mrs Perry's fallopian tubes were blocked, and therefore she was unable to conceive normally.

Mrs Perry and her husband disliked the idea of standard IVF, as it involves the creation of many embryos, some of which may be discarded. Their religious beliefs impelled them to seek a form of fertility treatment which would not involve the destruction of fertilized eggs. Their child was born from an egg which had been removed and frozen a few months before it was subsequently thawed, fertilized and implanted.

Egg cells are far larger than sperm cells, and contain a higher proportion of fluid. When frozen, this fluid can expand and damage the cell. Initially, the risk of such damage was deemed too risky to permit the implantation of embryos created from frozen eggs. However, the use of the technique in other countries seemed to demonstrate that babies born from 'frozen eggs' did not suffer ill effects. In 2000 therefore the HFEA decided to allow British women to avail themselves of this option.

Egg freezing was initially envisaged simply as a way of overcoming infertility resulting from treatment for cancer within the parameters of conventional IVF. Frozen and thawed sperm was routinely used in treatment so the possibility of using eggs in the same way was not expected to raise any additional questions.

The older mother

Remedying infertility caused by 'authentic' medical conditions may seem unproblematic, especially when the patient is a married woman of reproductive age, such as Mrs Perry. However, the availability of egg-freezing technology has opened a whole new realm of possibilities.

Rising infertility rates are frequently attributed to 'lifestyle' factors. Women are marrying or settling down with permanent partners later in life. Having pursued careers and used contraception until they reach their mid-30s, they may only then seek to start a family. Another issue is the increasing divorce rate, as couples who already have children by previous partners may find that they wish to start families with their new partners. The sexually transmitted disease chlamydia, which can be symptomless, has also been blamed for decreases in fertility.[8]

One approach to the problem of rising infertility has been to urge women to find partners and commence childbearing at an earlier age.[9] This could be bolstered by a widespread campaign against the behaviour that gives rise to sexually transmitted diseases.[10] Unfortunately, these restrictive messages are not necessarily heeded by women, nor are they welcomed by feminists. Women have endured bitter struggles to achieve the right to pursue their careers, and to be liberated from the burden of unwanted early pregnancies. To be told that they must renounce some of these benefits if they wish to become mothers is a harsh demand, and one which will in any case come too late for women who are already in their 30s or 40s and experiencing fertility problems.

Some commentators who focus on late motherhood and female promiscuity seem to endorse a degree of blame for women who indulge in these behaviours and who find that they are infertile.[11] The implication is that they have brought their fertility problems on themselves, and should accept the consequences of having behaved in an unladylike manner. However, censuring women's behaviour in this way omits to take into account the ways in which male behaviour and social trends in general affect the population's reproductive capacities. Men's smoking[12] and drinking habits are believed to have an adverse effect on sperm production, as are obesity, stress and other lifestyle factors.[13]

Partnerships formed in the prime reproductive years for men and women often involve long childless cohabitations. When such a partnership breaks down, the woman may find herself past her reproductive peak and single. For men, the breakdown of such a relationship does not necessarily spell the end of their reproductive potential. It is still far more socially acceptable for a man to have a relationship with a younger woman than for an older woman to have a similar relationship with a younger man. A man of 50 may be regarded as being 'in his prime' whereas a woman of 50 is likely to be seen as 'past it' or over the hill. A woman who emerges from a long-term partnership is therefore at a double disadvantage: her prospects of finding another partner and of starting a family are significantly less than those of her former partner.

For all these reasons, to regard women as being solely responsible for the fertility problems which are emerging in our society is to consider only a small facet of the problem, and unfairly burden women with the consequences of what is a far broader issue.

This being the case, it is unlikely that an easy solution will be found to the problem of increasing maternal age, at least in the short term. Over the next decades there will be many women—and men—who find that their natural fertility is impaired, and who will therefore seek fertility treatment. To some degree, it could be said that such treatments are providing an alternative to early motherhood for women, while not obliging them to forego the benefits they have attained over the past decades.

Neverthelesss, fertility treatments are not guaranteed to work. Women produce fewer and fewer eggs as they get older. Eggs produced from the age of 30 onwards may be at increased risk of genetic anomalies, which can lead to miscarriage and birth defects. This is why egg freezing is of such interest. On one level egg freezing is simply another option for women suffering from medical infertility. On a broader level, it could offer women a solution to the dilemma of having to choose between careers and motherhood.

Women could have their eggs removed and frozen at the biologically optimal age for reproduction. The eggs could subsequently be fertilized and implanted at a time which would fit in with the woman's career and financial and social needs. As Dr Gillian Lockwood has said '... egg freezing may come to be seen as the ultimate kind of family planning'.[14] However, it should be noted that it is still very difficult to achieve pregnancy using frozen eggs, although further research may improve these figures.

Is it acceptable that women should adapt their biological and reproductive roles to their social, emotional or financial needs? The equalizing potential offered by reproductive technologies can evoke strong and passionate responses, as they challenge assumptions about the idea of motherhood. This is particularly evident in the discussions around older women who have children. The fact that some women can now extend their fertility in line with that of men has by no means been universally hailed as a desirable state of affairs.

Case 18: Liz Buttle

Liz Buttle, then aged 60, hit the headlines in 1997 when she gave birth to a baby following IVF treatment, becoming the UK's oldest mother.[15] Ms Buttle had apparently lied in order to gain access to the treatment, saying that she was 49.

The HFE Act does not place an age limit on who can receive treatment, but refers to the need to consider the welfare of the child. The HFEA Code leaves the interpretation of this requirement to the discretion of clinicians in consultation with their patients. While some clinicians feel that advanced maternal age is a factor which should militate against offering treatment to older women, others may feel that to do so is discriminatory, since there are no barriers to men becoming fathers right up to the end of their lives—provided their potency also survives this long!

The case of Liz Buttle seems to point out a paradoxical social attitude towards parenthood. When pop star Rod Stewart announced he was going to become a father for the seventh time, also at the age of 60, he was given ample coverage in the media, and there was no moral outrage. Instead, the focus was on the pop star's joy at having a new son.[16] However, when Liz Buttle's 'happy event' was discussed, the tone was entirely different, focusing on anxieties for the child. George Monbiot has suggested that some of the condemnation arose from an underlying assumption that once women are no longer attractive to men, they should no longer be regarded as fit to be mothers.[17]

Concern for the child may also focus on potential damage from using 'aged' gametes. But these arguments are not always applied in a logical or consistent way. When Patricia Rashbrook gave birth to her fourth child at the age of 62, much attention was paid to the possible ill effects for the child. Scant notice was taken of her husband (aged 60 at the time of the birth) and the effects that his age might have on the child. This was an interesting omission given that Mrs Rashbrook used a donated egg, so she was not the genetic mother of the child, whereas her husband was the genetic father.

Some commentators object to the idea of older women receiving fertility treatment on the grounds that older mothers will not have the energy or stamina to care for their children. Again, since this attitude is not applied to older fathers, it seems that there may be some discrimination going on. What is more, if one argues that parents should fulfil certain age criteria in order to be justified in having children, where should we draw the line? Why restrict procreation on grounds of age alone? Perhaps we should also say that would-be parents who are below a specified socio-economic threshold should be denied the chance of having children? Or that working parents should not have children? Or perhaps that parents who plan to have only one child should be prevented since their offspring will be deprived of siblings . . . ?

While it may seem desirable that people should only have children in optimal circumstances, in practice it is very difficult, first, to determine what optimal circumstances really are, and secondly, to draw up limits that are both just and reasonable. Much of the time, it seems that when highly rhetorical arguments are made against older mothers, although they may appear to be based on concern for the offspring, closer inspection reveals something more like distaste or prejudice against the idea of an older mother at the root of the argument. If, for example, having a likely lifespan of only fifteen years makes it wrong to have a child, this should also be applied to people who might have a medical condition which reduces their lifespan, and perhaps also to firefighters, people in the army, or people who take part in extreme sports which may put them at greater risk of premature death.

Yet the fact that nature has always ended women's reproductive capacities at a far younger age than that of men cannot be ignored. It has been suggested that this is an evolutionary development to free women for grandmothering duties. It may be hard to establish the evolutionary value (if any) of natural phenomena such as the menopause. Generally, we do not have to question such facts, as they are simply part of the status quo. But as soon as we can override nature, it seems important to establish whether there is any purpose or point in the disparity between women's and men's reproductive lifespans. Finding answers is difficult, but this does not mean that these questions should therefore not be asked. While the evidence is still awaited, it may be reasonable to exercise caution, establishing pressing need on the part of the woman and absence of harm to the children before making treatments universally available.

There is another comment to be made on this case from a feminist perspective. The physical and psychological burdens of pregnancy and childbirth often take a heavy toll on women, and these risks increase with age.

Perhaps women such as Liz Buttle and Patricia Rashbrook are exposing themselves to unreasonable risks in trying to extend their reproductive lives in line with those of men. It has been suggested that many women who are currently of optimal childbearing age are consciously delaying having babies, assuming that IVF will help them out.[18] Should we be concerned about this, if it is true? IVF is expensive and time-consuming. It imposes risks on mothers, and the chances of success for women aged 40 to 42 are only 10.6%, as compared to 28.2% for women under 35.[19]

However, perhaps we should accept that this is the woman's choice. Undeniably, reproduction at a later age is risky. Equally, childbearing when young may also involve sacrifices which may affect a woman's future wellbeing. Neither option is ideal. Although older motherhood may come at a price, provided that women understand and accept these risks, then perhaps there is no further issue. On this view, any reproductive technology, notwithstanding its risks and uncertainties, may be seen as extending choice, and remedying inequalities.

Choice might be the obvious term to use here, but making reproductive choices is not as straightforward as it was thought to be in the 1960s. The advent of the contraceptive pill was hailed as a liberating revolution, allowing women to live their lives free of the burden of unwanted pregnancy. As things have turned out, women are neither free nor unencumbered. The reaction to Liz Buttle's use of reproductive technologies shows that women are not fully at liberty to use these techniques to bring their reproductive faculties in line with those of men. Social expectations still prevail, and Ms Buttle allegedly had to lie to gain access to the technology. On a broader level, women's career opportunities are still constrained, as are their reproductive lives. Technology does not always offer an easy answer to the reproductive dilemmas faced by women today.

It is perhaps an inescapable aspect of reproduction that it is likely to prove challenging to prospective parents, to disrupt their life patterns, and generally throw their assumptions into disarray. After all, however much control we gain, our biological functions are still unpredictable and will not necessarily be amenable to our conscious wishes. Given this, are reproductive technologies really beneficial to women in increasing their control and choice in respect to their fertility, or do they simply place more burdens on them? Choice itself can be stressful.[20] And as Liz Buttle's experience shows, reproductive choice is not simply open to anyone who might want to benefit from it. Social pressures may coerce some women into seeking to reproduce early, while other pressures inhibit older women from availing themselves of these opportunities, or castigate them for having done so. Reproductive technologies cannot in themselves rectify these pressures, nor equalize the asymmetry of nature's division of reproductive roles.

Risks associated with fertility treatment

Can we be sure that reproductive technologies really do benefit those they are designed to help? It is almost too simple to point out that most of the users of ART are women. It is women who bear the risks and burdens of treatments, and who will usually bear the blame and carry the stigma when treatment is deemed unnatural or unethical, as was the case for Patricia Rashbrook. There is the potential of very considerable income for those in the medical profession who can satisfy the supposedly overwhelming desire of women to be mothers,[21] and this may translate into an interest in perpetuating that desire, even where it may entail serious physical risks to the women involved. This interest may also push women into the precarious situation of becoming what has been termed 'moral pioneers'.[22]

Experimental and innovative reproductive technologies tend to arouse passionate responses from society at large, and in the media. To an extent, this media interest may be actively sought by clinicians who seek to publicize their willingness to provide controversial treatments: sometimes, perhaps, no publicity is bad publicity, especially if it prompts other women to come forward and pay high prices for treatment. But the women involved in these controversial procedures often suffer unwanted media intrusion and general condemnation. Even those who disagreed with Liz Buttle's choices may have sympathized with her feelings upon reading newspaper articles in which vitriolic judgements on her ability and motivation for motherhood were coupled with comments on her apparently aged, decrepit, and unattractive person.

The moral, physical, and psychological dangers of fertility treatment are necessarily borne primarily by women, rather than by the prospective fathers. It may be that these risks exceed what is reasonable or appropriate, or even that women are being harmed or exploited rather than benefited by these treatments.

Case 19: Temilola Akinbolagbe

In 2005, the BBC reported the case of a woman who had died as a result of fertility treatment.

Temilola Akinbolagbe had been receiving fertility treatment. As is normal with IVF patients, she was given drugs to stimulate her ovaries into producing eggs which could then be surgically extracted, fertilized, and reintroduced into her body.

However, two days after having started the course of treatment, things started to go wrong. While waiting at a bus stop, Mrs Akinbolagbe started to feel ill. She called a friend, who immediately summoned an ambulance. Mrs Akinbolagbe was rushed to hospital.

She suffered a massive heart attack, and was put on life support. It became clear that she was not going to recover and five days later, the machine was turned off.

Mrs Akinbolagbe had developed a condition known as ovarian hyperstimulation syndrome, a side-effect of the drugs given to women undergoing IVF. Most women who suffer from the condition experience relatively mild symptoms. However, very rarely, it can cause fatal complications.[23]

To answer the question of whether the risks to women are acceptable, let us examine first what are the medical (as opposed to emotional or social) risks involved in reproductive technologies. As the case above shows, there is a real, although very slight, danger of death from the drugs involved in ovarian stimulation. However, there are many other possible side-effects associated with these treatments. This is a list provided by the HFEA detailing the possible adverse effects of fertility drugs:

> hot flushes, mood swings, breast tenderness, insomnia, increased urination, heavy periods, weight gain, there may be a small increased risk of ovarian cancer with prolonged use of some drugs, stomach pains, sickness and nausea, and headaches, increased risk of multiple pregnancy (twins, triplets or more), allergic reactions, night sweats, vaginal dryness, changes in breast size, breakout of spots and acne, sore muscles ...[24]

To add insult to injury, women are expected to inject these drugs themselves! Although self-injection is probably welcomed by many women as it spares repeat visits to the clinic, it can nevertheless be a daunting prospect and some women find it unexpectedly painful.

Aside from the side-effects of the drugs themselves, other aspects of fertility treatments also carry risks. Some treatments carry an increased risk of ectopic pregnancy (where the placenta implants outside the uterus) which is extremely dangerous for the mother and almost invariably fatal for the embryo. Many techniques involve invasive procedures for the woman, some of which may involve general anaesthesia, with all its attendant dangers. And there is the not inconsiderable issue of time and money. Fertility treatments are costly in both respects, and while supportive partners or the State may contribute to the latter, the question of time off work can prove to be a serious problem for a woman's career and finances.

Time off work is not simply limited to the occasions that the woman is actually at the clinic (and in IVF treatment, timing is crucial—she *must* receive certain treatments at exactly the right time). Recovery time and additional sick leave may well be necessitated by the side-effects. In addition to the risks described above, women may experience emotional and psychological difficulties.[25] Some of these may well be exacerbated by drug reactions, but even without these hormonal triggers, the very process of undergoing fertility treatment is likely to be as emotionally draining as it is fraught with uncertainty.

This uncertainty is itself a serious issue with respect to women's welfare in the context of assisted reproductive therapies. While IVF and other reproductive technologies' successes tend to be reported with media fanfare and pictures of cherubic babies, their failures do not make such 'feel-good' reading. IVF as a medical procedure has a disappointingly low success rate: the bald fact is for every one hundred women who embark on an IVF cycle, fewer than twenty-five will succeed in bringing a child to term.[26]

Reproductive technologies do not simply 'liberate' women to have children whenever they wish to, since treatment may often prove futile. Moreover, women may not be truly free, in that their reproductive choices are heavily influenced by the societies they live in. To some extent, new technological possibilities could be regarded as imposing further burdens on women who are desperate to become mothers. It is a disturbing fact that a number of women undergoing fertility treatments refer to feeling that they 'had no choice' but to seek whatever means might offer a chance—however slight—of having a baby.[27] Each new technological development constitutes a further challenge for a woman who feels that she must exhaust every possibility.

Fertility treatments may be expensive, invasive, painful, humiliating and time-consuming. They also carry certain physical risks, as exemplified by the case of Temilola Akinbolagbe described above. And because of the necessary connection between the woman's body and the embryo, a woman may be obliged to undergo invasive and risky procedures even though it is her partner's fertility which is impaired. In effect, women may undergo the treatment for men's infertility. It has been said that new reproductive technology, eg ICSI, may be described as a means for affluent men to father genetically related children by applying risky technology to women.[28]

There have been cases of women coming to Britain for fertility treatment in circumstances where the pressures and obligations on them to have a child (often a son in particular) are almost overwhelming. Would-be fathers may be unable or unwilling to understand that infertility is not inevitably the 'fault' of the woman. Women from some cultures may fear divorce, and consequent penury as well as social stigma, if they do not fulfil what is expected of them.

These scenarios are certainly on the extreme end of the scale, but to an extent these pressures, fears, and social burdens affect all women seeking treatment, who may put themselves through medical procedures and experiences that they would not otherwise countenance. Does this tend to instrumentalize women? Perhaps so. Even very successful and compassionate clinicians may create the impression that women's bodies are objects to be monitored and regulated. It is as if they are saying that if those bodies are treated with the right drugs and by the right people they will produce what they were supposed to produce.

Perhaps clinicians themselves sometimes have difficulty coming to terms with their own failures to achieve successful treatment, all the more so when these feelings are mixed with compassion for a patient who longs for a child. This can add further pressure to women who are eager not to disappoint the doctor. If doctors are too willing to empathise with desperate patients, the message they may unwittingly be sending is 'don't give up', encouraging the patient to put herself through yet more arduous treatment.

But for many women, such treatment—whatever its risks and uncertainties—may be their only chance of having a genetically related child. Given the social and psychological drive to have children, perhaps it is not surprising that the dangers, discomforts and low success rates do not dissuade people from pursuing these therapies. After all, most 'normal' pregnancies also impose considerable physical, emotional, and financial risk on women. Seemingly, having children at all is a precarious enterprise. A woman who is prepared to put her body through nine months of pregnancy followed by childbirth may feel that it is not so much more unreasonable to inflict on herself the ordeal of fertility treatment.

We may acknowledge that women's autonomy is limited in situations where their own desires and social pressures shape their decisions, but attempting to protect women from themselves might seem patronizing. All human beings are subject to social and other pressures. The truly objective, rational, and autonomous individual may be an unrealistic ideal. However, does this mean that we should facilitate women's reproductive decisions whatever their motivations, and regardless of the probable outcome?

Women's autonomy and fertility treatment

In the past, the law has had little regard for the autonomy of female citizens, whose desires and aspirations were supposed to be aligned with those of their male protectors. This legal refusal to recognize female autonomy was precisely what many feminists have fought against. The law cannot change every social factor which affects women's lives, but having accepted this, it still has a role to play in safeguarding women's autonomy specifically in the complex and developing field of reproductive technology.

There are a number of legal frameworks which may be regarded as protecting women's interests in the field of ART: the HFE Act, the general laws of the UK, and recent human rights laws. Statutory regulation requires that doctors and scientists inform patients as to the options and risks involved in their treatment, and then abide by the choices that these patients have made. But should feminists require any more than that? Perhaps women should be demanding more positive rights of access to treatment.

Autonomy can be respected in one or both of two ways in the arena of reproduction.

First, one can respect reproductive autonomy by abstaining from physical intervention in women's reproductive lives, for example by not forcing women to use contraceptives, to have sex, to undergo abortions, or to have caesarean sections during childbirth. This might be termed a negative approach: the focus is on refraining from physical interventions on women against their will.

Secondly, autonomy can be respected in terms of facilitating women's reproductive choices, perhaps by providing access to contraception, to abortion, or caesarean sections, or to fertility treatments as and when a woman may desire. This could be termed a positive approach: it requires positive action from those who are deemed responsible for meeting women's reproductive demands.

Intuitively, respecting the first kind of autonomy might seem more important than the second. However, it is sometimes argued that since there are no rules and regulations which restrict fertile people from conceiving and bearing children—even in the most inauspicious circumstances—it is unfairly discriminatory to apply rules and restrictions to the infertile. If this principle is accepted, it seems to undermine the validity of regulatory limits.

Where access to fertility treatments, contraceptive, or abortion are restricted, this does indicate that women do not have the right to choose freely their reproductive destiny: they must choose within the parameters that the regulators dictate. Some people resent this constraint, and consider that in fact it is none of the State's business to intervene in women's reproductive choices whether by physical coercion, or by legislation which restricts access to fertility services.[29] Full reproductive autonomy on this view might include unlimited access to whatever fertility treatments are technologically feasible.

There are several problems associated with this argument. Most obviously, preventing fertile women from conceiving or bearing children necessarily involves physical coercion. This is widely held to be unjustifiable in almost any circumstances, even if non-intervention would inevitably lead to the offspring and/or the mother suffering severe damage or even death.

The other point to make in this context is that 'natural' conception is not by any means completely unregulated. On the contrary, it is the hallmark of human civilizations that restrictions and laws are introduced in relation to, for example, the age of consent and marriage, rape, abortion, adoption, and other aspects of reproductive law. This might be taken to reflect the inherently social nature of reproduction. Having children cannot always be assumed to be an independent or individual endeavour; neither is it necessarily a private matter between consenting parties.

It seems clear that the law should seek to uphold and protect women's bodily autonomy just as much as it does men's. Neither men nor women should be subjected to procedures without their consent either for their own benefit, or for someone else's. However, this does not answer the question of whether women should have unlimited choice when they seek treatment. As suggested, the primary reason for non-intervention in fertile women's reproductive choices is the fact that this would involve physical coercion. These decisions may still be regarded as reprehensible, immoral, or foolhardy; we do not lose our capacity to make moral or legal judgements concerning reproductive decisions just because we are compelled to refrain from physical intervention.

In the case of reproductive technologies, judgements can be made without this necessarily involving physical coercion of the women involved, and this means that restrictions can be set up which prevent access to certain treatments based on their cost, efficacy, risk to the mother, or danger to the child who may be born. Reproductive cloning, for example, might fall into all of these categories.

So what should happen if any of these considerations, whether the mother's own wellbeing, the wellbeing of the future child, or other social or economic factors, seem to weigh against a prospective mother's reproductive choice *and* that prospective mother requires fertility treamtent in order to achieve that choice? As suggested, reproduction is not simply a private phenomenon; throughout history, societies have taken the greatest interest in who reproduces, and how. Provided that this interest does not veer into physical coercion, it is not necessarily intrusive or unreasonable. Nor is it necessarily unjustly discriminatory.

The question of what grounds justify non-provision of fertility treatment, however, is a vexed issue. Respect for women's reproductive decisional autonomy does not guarantee them access to reproductive technologies in all circumstances. Even so, the means by which we attempt to delineate the kinds of circumstance in which fertility treatments can be offered are extremely controversial.

The present situation may not be ideal, in that clinicians are required to make judgements concerning a woman's suitability to become a mother. Changing social values alter perceptions of what is necessary for family life and for children to flourish. Assumptions about the type of person who merits treatment may vary from clinic to clinic, and sometimes decisions may be unjust or based on prejudice. It is the job of the HFEA to ensure that decisions are made openly, honestly, and consistently.

There are limits to women's autonomy in this respect just as there are limits to all adults' autonomy in terms of what they can demand. So, is the autonomy of women adequately protected under the law and practice of reproductive technology? Such laws as exist are drawn up to uphold women's interests in maintaining control over their bodies. But in terms of guaranteeing their right to access treatment, the answer is less clear.

Regulation and law can only go so far in influencing public attitudes. Women's autonomy may be undermined by popular perceptions of women as mothers, or as wanting to be mothers at any cost. This is reinforced by the press ('my miracle baby'), to the fascination of the warm-hearted public. Infertility is increasingly portrayed as a disease for which there is a cure if the woman tries hard enough and spends enough money.

The law and the regulatory environment around reproduction and ART are designed to ensure that women's interests are not subsumed by their own desires or those of society. However, as we have pointed out, the law is a blunt tool if it is to be used for this purpose. The subtler, and perhaps more powerful, influence of culture and social expectation are inextricably involved in women's reproductive choices and desires.

Since these social factors are constantly changing and evolving, the regulatory environment needs to maintain scope for flexibility and responsiveness. Perhaps the best way to achieve this is to encourage and stimulate debate on the subject. Openness and transparency on the part of the regulators and a willingness to respond to challenges will be needed to ensure that the benefits of good regulation serve to enhance and protect women's interests without becoming rigid or paternalistic.

Meanwhile, we need to continue to question whether it is being assumed that all women are incomplete if they have no children, whether voluntarily or involuntarily. There is a risk that new techniques, which have scarcely been tested as safe or viable, are rushed into use by clinicians in order to enable them to be ahead of the field and advertise themselves as such. Prospective parents may re-mortgage their homes and dispose of their life savings in order to undertake repeated attempts at fertility treatment, sometimes even in the absence of a conclusive diagnosis.

Women are not simply defined by their biological functions. Gender roles, including parental responsibilities, are hugely influenced by social factors, some of which seem more intractable even than the technical challenges which have perplexed our scientists. The greatest advances in reproductive technologies have the potential to become factors in the suppression of women's autonomy and integrity if they are mismanaged, injudiciously regulated, or if their very availability serves to further prejudicial attitudes and assumptions.

It is necessary to remember when considering the advance of new reproductive practices, that it is important that every woman, like every man, should feel like a worthwhile and whole person whether or not she has children. Childless women must not feel that they have no choice but to subject themselves to every new reproductive technique that is developed. We as a society need to ensure that we do not perpetuate the idea that a woman's life is thwarted, embittered or blighted if she does not become a mother. While the law can seek to ensure that the ugliest of infringements on women's autonomy and bodily integrity are not permitted, it is the responsibility of all women to try to ensure that their expectations do not drive them to relinquish that very choice which women have fought so hard to obtain.

Future developments

What do the reproductive technologies of the future hold in store for women?

There are many challenging possibilities on the horizon, with the potential for far-reaching effects on women's reproductive lives, and on the imbalance in reproductive faculties between men and women.

Artificial gametes

The first of these is the development of artificial gametes. There are several techniques currently being explored by scientists.[30] In certain culture conditions, stem cells from mice can differentiate into sperm and eggs. If applied to human beings, this might mean that scientists could cultivate sperm and eggs from human embryos. Combining this with the technique of therapeutic cloning, it might be possible to produce gametes for people who are unable to produce them in their own bodies.

This could work by taking a cell from the infertile adult, for example a skin cell, and inserting its nucleus into an empty egg cell.[31] This is the same process involved in cloning, but in this case, the development of the embryo would be halted in order to form stem cells. These could then be treated with chemicals to stimulate them into becoming sperm or egg cells.

Eggs and sperm obtained in this way would be very similar to the ones which the adult would have produced if fertile, although the process would be likely to be difficult, time-cosuming and costly. But if the technique could be applied to human beings effectively, it might circumvent many of the problems associated with donated gametes. Currently, parents who conceive children after fertilization with donor gametes must face the difficult problem of deciding how and whether to inform their offspring of their background. Donors must face the fact that any children born from donated gametes will have the right to find out the donors' identity.

If there is an alternative to involving a third party, it is likely that this will be eagerly sought by prospective parents, especially if this also means that it is their genes, rather than a stranger's, which will be transmitted to the offspring. But the prospect of artificial gametes also raises some interesting questions. As well as enabling infertile people to produce gametes, it could theoretically be used by people who are not regarded as being infertile. It has been suggested that same-sex couples could use the technique to have children genetically related to both partners.[32] Post-menopausal women would be able to produce eggs up till the end of their lives. Taken to the extreme, single people might be able to use the procedure to generate sperm or eggs in order to fertilize their own gametes, enabling them to reproduce without a partner.

These possibilities are likely to re-ignite debates about who should be able to reproduce and how. They may also raise issues about funding. Currently, the provision of NHS-funded fertility treatment is patchy in the UK. Many people believe there should be more funding of IVF and related techniques, and that this should not be restricted on the basis of sexuality or other 'social' judgements. If the availability of artificial gametes became a reality, the number of claimants coming forward for fertility treatments could be hugely increased, putting a strain on resources that are already stretched to breaking point.

From a feminist perspective, the prospect of creating gametes outside the body might be beneficial to women. IVF involves drugs and surgery in order to harvest eggs. The possibility of using artificial gametes could free women from the burdens of egg-collection, as well as prolonging their fertility past the menopause. This might enable women to approach reproduction in a similar way to men. However, embryos conceived using artificial gametes would still need to be carried by a woman. If a single man, or two men wanted to have a child using this technique, they would need to use a surrogate mother.

Artificial wombs

This brings us to another exciting—if controversial—area of research. The possibility of incubating babies in artificial wombs has long been regarded as the province of science fiction. However it no longer seems incredible that, within the foreseeable future, babies could be gestated without the need for a woman's body.[33] Ectopic pregnancies show that the development and survival of a foetus outside the uterus is possible[34] and advances in neonatal care have meant that the number of weeks' gestation necessary for a baby's survival are diminishing steadily.[35] Meanwhile, at the other end of the spectrum, laboratory techniques have enabled scientists to create and cultivate embryos *in vitro* for up to two weeks before implantation. Restrictions on this time frame have been due to legal and ethical cut-off points rather than technical problems. Thus, the window of time required for pregnancy is shrinking, and could feasibly become redundant.[36]

A prospect such as this is likely to cause a high degree of controversy. But should we shy away from the possibility of relieving women from the burdens of gestation and childbirth? 'Pregnancy is barbaric'[37] said Shulamith Firestone, and perhaps she had a point. The evolution of the human species has resulted in babies' heads growing ever larger. Human beings have one of the most protracted, difficult and dangerous labours in the entire animal kingdom. Even with modern medicine and technology, many women suffer physically and mentally during the process of gestation and delivery in ways which can adversely affect their ability to function at work, and as mothers. Women die in childbirth even now, and many children suffer irreparable damage during birth.

Even with all the currently available technology, gestation of the foetus is still a woman's exclusive province. The more advanced scientific knowledge becomes, the more pressure is put on pregnant mothers, with recommendations on what to eat and drink, what medications to take, and what pre-natal diagnostic tests to undergo. Meanwhile, still more dramatic concerns for pregnant women are raised by the prospect of foetal surgery. Operations can now be performed on foetuses in their mothers' wombs.[38] When defects such as spina bifida are spotted on pre-natal scans, they may be rectified surgically before birth.[39] Naturally, this cannot be achieved without cutting open the pregnant woman, and imposing the risks of surgery on her too. However foetuses in the womb heal much quicker after surgery than newborn babies, and therefore mothers may feel obliged to undergo surgery in order to maximize their child's chance of a good recovery.

As imaging and diagnostic procedures become more sophisticated on the one hand, and foetal surgery techniques improve on the other, pre-natal surgical interventions are likely to become more common. This adds to the already burdensome nature of pregnancy. Given this, perhaps women should be relieved of the problems involved in pregnancy and childbirth altogether.

The implications of this for women could hardly be over-estimated. Throughout the history of mankind, women's lives have been determined by their biological make-up. Of course, there are women who are willing—even eager—to undergo pregnancy and natural childbirth. However, this should not cloud the issue. To return to a feminist point of view, it seems that there is an injustice in the fact that if a man wants to father a child he can do so without undergoing the pains and risks of gestation and childbirth. However, a woman cannot. If we can use technological means to redress this balance, why should we hesitate?

Perhaps one way to address this problem would be to allow those who want children to use surrogates. That way, women as well as men could reproduce without having physically to give birth. Is there anything wrong with this suggestion? Considering the role of the women acting as surrogate mothers, there might be something repugnant in the idea of a possibly vulnerable group in society being systematically used to enable another group to avoid having to go through undesirable physical experiences.

At the moment, if men want to have children, they are obliged to rely on a vulnerable group in our society (ie women) to act as surrogates, and undergo the risks and burdens involved. If it would be unjust for some women to impose this situation on certain other women, can it be just for society as a whole to expect it of all women on behalf of all men? Clearly, as long as no other option exists, it is unfair to blame men for being reliant on women in this way. And many women will undertake the reproductive enterprise for their own benefit, rather than on behalf of a man.

Nevertheless, it remains true that pregnancy and childbirth take a huge toll on women around the world. The chance of a technological solution would offer women genuine reproductive equality with men, and should not necessarily be dismissed out of hand.

The possibility that children would be damaged by not being physically inside their mother cannot be ignored. Within an artificial womb the developing foetus might lack some essential nutrients or environmental conditions. Equally, mothers and children might simply fail to bond if the natural physical link is severed. One possible response to the question of damage resulting from growth within an artificial womb is simply that the procedure should be tested and researched as thoroughly as possible before being used (although admittedly it is hard to see how this could be done).

While many—probably most—prospective mothers willingly curtail their lives to benefit their unborn children, there are many factors over which they have no control. Babies in artificial wombs, on the other hand, would be free from the harmful effects that a mother's stress can have on an unborn child. A mother's use of drugs, cigarettes or alcohol would no longer adversely affect the child's health. Pollution, car fumes, high maternal blood pressure, a heavy fall—all of these may damage or kill an unborn foetus. In the artificial womb, the child could develop peacefully in the absence of these dangers.

These benefits might outweigh any theoretical risks affecting children born through ectogenesis. But what of the maternal bonding issue? Perhaps children and mothers would simply fail to establish a connection if the child had not been carried under its mother's heart for nine months.

One point to make here is that in the current age of prenatal testing and diagnosis, the phenomenon of the 'tentative pregnancy' has been well-documented.[40,41] Women undergoing pre-natal tests sometimes speak of a need to re-form bonds with their foetuses after receiving test results. There are also considerations related to the development of visualization techniques during pregnancy: for many couples, being able to 'see' the baby on screen is a highly significant moment. This is not connected with the knowledge that the baby is inside, which has been evident all along, but is tied up with the idea that the baby looks like something, that it can be seen despite being inside the mother.[42]

Perhaps the strongest argument against the idea that mothers have to gestate their children in order to bond with them is the existence of fathers. While fathers may sometimes get a bad press, there is little evidence that the mere fact of being a man, and therefore not gestating the child, makes men unable to love or bond with their children. Moreover, parents who adopt children might well feel affronted at the idea that however much they may love their children, they are somehow lacking an essential connection with them.

So what would it mean for women if artificial wombs became a reality? For the first time, women would be equal parenting partners with men, neither more nor less responsible. Parenting decisions could be made independently of physical constraints. An idealized concept of motherhood might be lost, but women might need to weigh this up against what they could gain. As we have seen, much of the supposed choice that reproductive technologies have given women is subverted by social pressures, as well as by the fact that gestation and childbirth are still inextricable from women's bodies. Ectogenesis might liberate women from the latter, but social unease still may render it unacceptable. Whatever the answers to these questions, it seems evident that interesting times lie ahead.

CHAPTER 5
Private Lives and Public Policy —The Story of Diane Blood

One of the most fraught issues relating to reproductive technology is the extreme emotions which can be experienced by people seeking treatment. Often, the plight of an individual may be deeply moving, and it is easy to become caught up in the emotive appeal of a particular case, especially when these aspects are seized upon by the media. However, the HFEA is obliged to consider its regulatory decisions from a broad perspective, encompassing considerations beyond the needs and desires of specific individuals. The creation of a precedent can have far-reaching effects on the rest of society. Where, viewed in isolation, an individual's case may be compelling, these individual emotions play an invidious part in influencing decisions which affect the whole of society.

This chapter explores the tensions between individual desires and the need for public accountability and coherence in law. It also addresses the problems which can arise when the emotive issues related to a particular dilemma are taken up by the media, with the danger that these can put intolerable pressure on the regulatory body and in the worst cases, can forestall the possibility of reaching a reasoned decision.

In previous chapters the discussion has been illustrated with a number of relevant cases. With reference to the tension between public and private, and moral responsibility and the media, one particular case presents itself as having overwhelming resonance in both these areas. This is the case of Diane Blood.

Case 20: Diane Blood[1]

In 1995, tragedy struck in the life of a woman named Diane Blood, an advertising executive. Her husband, Stephen, had contracted meningitis, and lapsed into a coma. When it became clear that he was not going to recover, Mrs Blood asked doctors to retrieve some of her comatose husband's sperm so that she would be able to have his child even after his death.

Fortuitously, a doctor who was an expert in the field, happened to be at the hospital at this time. However, Mrs Blood was told that there might be legal problems over obtaining the sperm. Nevertheless, the decision was made to go ahead. A sample of semen was obtained by electro-ejaculation without Mr Blood ever recovering consciousness. The life support machine was then turned off. A second sample was taken and Mr Blood was declared clinically dead. Mrs Blood was distraught at her husband's death, but she believed that at least she could still bear his child. Yet when she took steps to obtain the sperm from the hospital, she received some unexpected information: the sperm had been retrieved illegally, and the HFEA was refusing to allow her to use it.

According to HFEA regulations and medical law gametes cannot lawfully be removed or stored without the written consent of the person from whom they were obtained. Since Stephen Blood had been in a coma, he had been unable to consent either to the sperm retrieval, or to its storage. He never knew what had taken place. It was therefore unlawful to continue to store the sperm, and Mrs Blood was informed that she would not be allowed to use it for treatment either.

Meanwhile, the media, sympathetic to Mrs Blood's case, took up cudgels on her behalf, and the HFEA was portrayed as a bureaucratic and unsympathetic machine with no regard for the heart-rending circumstances in which Mrs Blood found herself.

In fact, Mrs Blood eventually won the right to take her husband's sperm abroad. She was successfully inseminated with it, and to date has had two children posthumously by her husband. The law was clarified in the wake of this case, and it remains illegal to remove, store, or use a person's gametes in this country without that person's written consent.

Many British people will recognize the name 'Diane Blood'. The case, as outlined above, shows why her plight attracted so much attention. There is no doubt that the premature death of her husband was tragic for Mrs Blood, and this tragedy was compounded by the knowledge that although her husband's sperm was being stored, she was legally unable to use it to have the child she longed for.

So what were the HFEA's reasons for restricting Mrs Blood's options in this case? Should Mrs Blood's private decision to obtain and use her husband's sperm have become a matter for public debate in this way? This is a case in which the tension between private reproductive decisions, public interest, and regulation was brought sharply into focus.

The interest and appeal of Mrs Blood's position do not perhaps need much additional exposition here; her motivation and decisions speak for themselves. It was clear that she simply wished to have her husband's children at all costs, despite the fact that the children would never know their father. In fact, these desires are not uncommon immediately after the death of a spouse. However anecdotal evidence suggests that, in many cases, over time childless widows tend to feel glad that, they did not have the opportunity to satisfy this longing, which is often gradually replaced by a preference to have a child with a father present.

While Mrs Blood's motivations were straightforward enough to appeal to the media and the public, those of the HFEA were less obvious. Mrs Blood was clearly the victim of a terrible tragedy. Why did the regulatory body feel the need to compound this by refusing her access to the sperm which had been stored?

Consent

One of the fundamental features of the HFE Act is the emphasis that it places on consent. Consent is not only a critical concern in the medical world, but is also a vital consideration in the area of sex and reproduction. As discussed in Chapter 2, the requirement that doctors obtain consent before carrying out surgical procedures is regarded as one of the foundations of medical ethics.

The law requiring consent is framed so as to protect our bodies from unwanted and unwarranted interventions. The reasoning for this is so obvious as to leave little need for explanation. Our bodies should not be regarded as the property of others. However, when we are dying or unconscious the possibility of giving consent for medical interventions may disappear, and if the strictest observation of consent requirements were adhered to, unconscious patients could not receive lifesaving emergency treatment. Clearly this would not be desirable and a balance must be struck.

The law stipulates that if someone is unconscious and unable to give consent, medical interventions may be permitted if they are urgently required to save the patient's life or health. Likewise, the Council of Europe Convention on Human Rights and Biomedicine 1997, Article 8, states: 'when because of an emergency situation the appropriate consent cannot be obtained, any medically necessary intervention may be carried out immediately *for the benefit of the health of the individual concerned*' [our emphasis]. In all other medical circumstances, all common law and European principles insist on informed consent.

When a patient is unconscious, then he or she can be treated, but only if this is deemed to be for the benefit of the patient's health. Non-urgent treatment will have to wait until doctors are able to consult with the patient and obtain consent.

Contrary to popular belief, next of kin and spouses are not legally entitled to give consent for medical interventions on behalf of a loved one who is unconscious. Consent can legally be given on behalf of someone else only if the person giving consent is the parent or legal guardian of a minor. Thus, since Mr Blood was not a minor, and the removal of sperm was not necessary to extend life or preserve health, it seems evident that plans to remove sperm should have waited until he had recovered sufficiently to understand and sanction what was being done to him. If a patient never regains this capacity, the patient's sperm (or eggs) must simply remain un-harvested.

Two arguments can be brought against this. One is based on the idea that since a dead or dying patient no longer has interests, it does not matter what is done to that person. The second suggests that interventions in a patient's 'best interests' should be permitted, where such interests are understood to encompass more than clinical outcomes.

Does it matter what happens when we are irretrievably unconscious or dead?

If it is assumed that a dead or dying patient can no longer benefit from his or her body parts, why should those body parts be 'wasted', rather than automatically removing them and redistributing them to others who might receive some genuine benefit from them? In essence, this argument stems from a conviction that the dead no longer have any interests, or that if they do, the interests of the living are greater, and can override them. There is something to be said for this view, but it is clear that in practice it is not consistent with normal attitudes towards people before and after death.

In any event, we are talking here about a patient who was not yet dead when the first sample of his sperm was removed. Mr Blood was on a life support machine which was sustaining his vital functions. For a patient in these circumstances, the chances of full recovery may seem remote. However, it is also clear that any chance of recovery rests on the continuation of life support for a patient who is unable to sustain his vital functions unaided. During this period of comatose life, there might well be medical procedures which could benefit him in the narrowest sense of prolonging his life. So for example, if a comatose or dying patient suffers a ruptured appendix, or develops pneumonia, he could receive treatment for these conditions despite being unable to consent.

As long as a patient is not yet dead it is possible to make a distinction between clinical interventions which may improve his medical condition and prognosis and those which do not. It is quite evident that the removal of sperm at the request of a third party cannot improve the comatose person's medical condition, nor extend his lifespan. In essence, it is not an intervention for the clinical benefit of the health of the dying person, and therefore inasmuch as the dying patient can be said to have interests, it is not in those interests to have sperm removed.

We might admit that sperm should not be taken from comatose or dying individuals without their prior consent, but surely after they are dead such a procedure can do no harm? This question requires us to look in more depth at the idea that we have no interests after we are dead.

Many of us, whether religious or not, have a preference as to whether we should be buried or cremated. (Indeed, Mrs Blood herself has written that her husband had specifically said that he did not wish to be buried, and that she herself had a 'horror of being cremated'.)[2] Concerns about what happens to us after our death are not out of the ordinary. Some of us carry donor cards indicating that we wish our organs to be available to others after our deaths. Those who do so may specify which organs they would be willing to donate. Bodies can also be 'bequeathed' to medical science, or to 'artists' such as Gunther von Hagens. Even so, many people would feel horrified at the idea of their bodies being on public display, or being dissected in this way.

Whether or not it is correct to say that we cannot be harmed or benefited after our death, there is still a tendency to act as though we have interests in what happens to ourselves, our bodies, and our property after we die. The law in this country recognizes and respects wishes of this kind, regardless of their logical coherence. There is a universal custom, respected by non-believers as well as the religious, of elaborate respect for the body in the time between death and burial or cremation. There is a similarly strong presumption that where a person has recently died, that person's wishes concerning his or her body and property should be upheld. In order to ensure this, people are encouraged to provide adequate evidence of their wishes by making a will.

There is also an assumption that some things are permissible in relation to dead bodies, and some are not. So, for example, in most societies eating the flesh of a dead person is not regarded as permissible even if it cannot be regarded as 'harming' that individual per se.

The idea that people have interests in controlling what happens after they die means that their wills are rigorously adhered to. This presumption is strong enough to entail that even if a will seems vastly unfair (for example, if a man leaves his fortune to the lavish upkeep of his pet dog instead of the impoverished sister who devotedly nursed him during his illness), it would not automatically be overridden. Rational or irrational, we are disposed to act as though the dead do have some interests after they die, and that—all other things being equal—we should comply with their wishes. If we did not hold this belief, we would not write wills, and the assets—and organs—of dead people would simply be distributed in whatever way seemed most appropriate.

Informed consent

If we do afford respect for the wishes of people after their death, or in circumstances in which they are irrevocably unconscious, it is relevant to consider what the person in question might have understood by requesting a particular procedure, or action. Many of us, even though we accept that we cannot feel pain or humiliation after death, still hope that our bodies will not be tampered with. We hope that we will be treated with dignity and respect.

The means employed in putting our bodies to use are of vital importance when we decide what we want to happen to us after we die. For example, some people hate the idea of being cremated. Equally, some people who carry organ donor cards specify that they do not wish their eyes to be removed after death—they may be able to countenance the removal of internal organs, but the thought of their eyes being extracted is too painful to contemplate.

In the case of Mr Blood, the method by which his sperm was extracted is an important consideration. Few people are likely to be fully aware of the procedures involved in sperm retrieval from the comatose or dying. Obtaining sperm or eggs is hedged about with strong social, moral, and religious taboos relating to sex and bodily integrity. These taboos may affect someone's decision about the acceptability of having his or her own sperm or eggs harvested. Such concerns might well turn on the specific methods employed.

The technique used to extract sperm from Mr Blood is known as electro-ejaculation. This involves the insertion of an electric baton into the patient's rectum, with the electrodes close to the prostate gland. Electrical stimulation is administered rhythmically and the voltage is progressively increased with each stimulation until ejaculation occurs. This usually results in what is called retrograde ejaculation, meaning that the sperm are passed into the bladder. The baton is then withdrawn from the patient's rectum. Sperm is subsequently collected from the bladder via a catheter.

It is an open question how much information is needed in order to provide informed consent. But it seems undeniable that misunderstandings or incomprehension about what a procedure involves are likely to compromise the validity of consent in some circumstances.

A patient may consent to an intervention, and the intervention as described may entail or bring about certain conditions; however, the patient may not consent to those conditions because he or she may not grasp the entailment relation or the causal connection.[3]

On this note, it is worth bearing in mind the widespread dismay in the UK when the story broke that several hospitals had stored tissue taken from babies who had died. Parents' consent had been obtained, but this had been based on a misunderstanding of the kind and quantity of tissue to be removed and retained. This was deemed to mean that their consent had not been properly informed, and was therefore invalid.

For some people, the mechanism by which sperm is collected may seem irrelevant. Mrs Blood herself mentions in her book that the doctors did not tell her much about the procedure, and the implication is that she did not want to know. However for some people the prospect of their bodies being invaded in this way might be regarded as being tantamount to a sexual assault, and the way in which one interprets this is extremely subjective.

Even if it could have been established that Mr Blood had agreed to the collection of his sperm, his consent would not necessarily have been valid unless he had been fully aware that it would entail the procedure described above.

Dignity

When the Blood case was in the news, those who were uneasy at what had been done to Mr Blood sometimes referred to the lack of dignity involved in the procedure. However, Mrs Blood herself has pointed out that this criticism rests on an assumption that, left alone, dying patients do have dignity. Poignantly, she describes the utter lack of dignity involved in the process of death which her husband suffered. Undergoing further interventions such as that used to retrieve his sperm may have seemed irrelevant in the context of the total lack of dignity and control suffered by the dying patient.

Dignity is a concept which is notoriously hard to pin down.[4] However, it is often associated with respect for autonomy and with integrity, both physical and moral. Patients whose ability to make their own decisions, or to control their bodily functions, is impaired may feel this acutely as a loss of dignity. Patients who have lost all or most of their faculties may be too ill to experience their loss of dignity as such. But this does not stop us dreading and fearing the day when it may happen to us.

Dignity may also be violated where the wishes and desires of other people are forcibly imposed on a person's body. When prisoners in Iraq were forced to adopt sexual postures, this was a deliberate attack on their dignity.

Mrs. Blood's claim that her husband's death was not a dignified process seems plausible. However submitting to the encroachments of death or illness is a separate matter from being acted upon by other human beings.

At the very least it seems true to say that no individual was responsible for inflicting upon Mr Blood those indignities which are involved in dying from meningitis. No-one had chosen to put him through this ordeal. However, the retrieval of sperm was not caused by the inexorable progression of a devastating disease. Rather, it was the exercise of one human being's will over another human being's body.

It is not possible now to establish whether Mr Blood would have thought that electro-ejaculation would have been an affront to his dignity. Certainly there are many people who would dislike the idea of such a procedure being carried out on themselves. In the absence of such objective answers, again, it seems important to try to ensure that interventions are not carried out without evidence that this is what the patient would have wanted.

While it may be illogical to believe that the dead have interests, or that we should honour the wishes of dead people, much of our society is premised on the assumption that people do have such interests. If an exception were to be made in the case of Mr Blood, perhaps this would lead to the conclusion that the wishes expressed in wills need not be respected, and that dead bodies should be subject to whatever acts and procedures might be advantageous or enjoyable to others, perhaps extending even to cannibalism or necrophilia. Unless we as a society can countenance this prospect, we need to accept that there is some degree to which the dead or dying do have interests, and that—all other things being equal—it is appropriate to respect these interests.

Is there a doctrine of 'best interests'?

The second argument in favour of obtaining gametes without consent recognizes that we do give some credence to the idea that people may have interests even after death, and that certain actions may be taken to further these interests. As discussed, medical interventions can be performed on patients without their consent, but only if this is deemed to be lifesaving. Treatment of the mentally incompetent, however, may be undertaken on the basis of a broader understanding of their best interests. It has been suggested that the same principle should be applied to the comatose prior to certain death. Perhaps it *was* in Mr Blood's best interests for his wife to be able to conceive a child.

But while there is—in theory at least—an objective answer to what is in someone's interests in the narrow life-saving sense, the question of broader interests does not admit any easy answers. A judgement about a patient's clinical interests can be based on evidence and experience. It can be supported with reference to statistics and probability, and can be validated by colleagues and settled on the basis of shared facts and expertise.

Clinical and other interests

It is essential to recognize that the kind of narrow clinical interests which doctors take into account are not the whole story. However, other interests are far harder to determine and more subjective. In the broader sense, 'interests' will encompass thoughts, beliefs, aspirations, perhaps religious convictions, and many other factors which make up an individual's conception of what constitutes a 'good' life.

Some people might regard fathering offspring as being desirable, while others might not. As long as patients are in a position to be able to articulate their concerns and interests, these difficulties are not insurmountable. However, if a patient cannot articulate his or her wishes, how are we to know what really is in the patient's best interests in this broad sense?

The answer could lie in an appraisal of what is good for people: we could simply assume, for example, that it is objectively in someone's best interests to have offspring. However, it is questionable who would be entitled to make these judgements.

Perhaps turning to the doctor might prove helpful here. Doctors are likely to have seen many such situations, and may have useful experience in understanding how to deal with them. But to assume that doctors are in a position to be able to weigh up these different possibilities and arrive at the 'right' conclusion seems misguided. Doctors, after all, are trained in medical skills and clinical judgements. There is no obvious reason why they should be arbiters of more subjective abstract or moral judgements, any more than other members of society.

Mr Blood might have felt, had he foreseen the circumstances, that his sperm should be retrieved. Equally, some people abhor the idea of bringing children into the world if they would be unable to bring them up, and some strongly believe that bringing extra children into an overpopulated world is immoral. Even if we admit that having children could be in the interests of a patient in Mr Blood's situation, the difficulty of establishing that this is so in this particular case remains.

Parents do this as a matter of course for children who are deemed to be too young to recognize or act in their own best interests. Parents are generally deemed to be the best judges of what is good for their children, and this does not necessarily strike us as unreasonable. As adults, however, the idea that someone else may be better placed to know what is in our own interests than we are ourselves, is unappealing. In the medical context, this is known as medical paternalism, and has been associated with some unpalatable practices. Doctors who wish to act in a patient's best interests, and who do not regard the patient as being the appropriate person to identify his or her best interests, may decide to perform procedures without the patient's consent, or even against the patient's express wishes.

An earlier chapter of this book highlighted the case of a doctor who—in a classroom full of students—anaesthetized a female patient with chloroform and impregnated her with the sperm of one of his students without her knowledge or consent. To our modern minds, this procedure may seem utterly appalling, and in the modern legal framework the doctor would certainly have been liable to a charge of assault. However, it is likely that the doctor concerned believed that he was acting in the best interests of the patient. Is this kind of scenario really one to which we would be willing to return?

Conflicts do sometimes arise where a doctor feels that a patient's idea of his or her best interests is grossly wrong.

Case 21: MS

In 1996 a woman attended a local doctor's surgery in order to register as a patient. MS was 36 weeks pregnant.[5] On examination, it was revealed that MS was suffering from a condition known as pre-eclampsia. This is highly dangerous both to mother and foetus.

The doctor advised MS that her baby should be delivered immediately as a matter of urgency. MS refused, saying that she wanted the child to be born naturally. The doctor suspected that MS was depressed and was concerned that she did not comprehend the gravity of her situation.

However, MS was a qualified veterinary nurse. She had a full understanding of the condition from which she was suffering, the nature of the treatment proposed, and the possible results of 'allowing nature to take its course'.

Despite this, the decision was taken to dispense with the need for her consent in this case. She was operated on against her wishes.

MS took her case to court. It was determined that she had been unlawfully detained under the Mental Health Act in order to override her lack of consent, and that her autonomy could not legally be overridden even though she might have been thought to be acting irrationally.

This case shows the dangers involved in allowing medical professionals to dictate to patients what is in their best interests. The ruling asserted that the apparently irrational desire of a patient to refuse a caesarean must be respected.

These situations are likely to be terribly frustrating for the medical professionals involved. Nevertheless, there is a vital principle to be upheld. Doctors' medical training does not necessarily equip them to interpret the nuances of patients' beliefs and value systems, to impose their own values and judgements, or to forcibly impose treatment on someone who does not consent. The ruling here is interesting in the context of what happened to Mr Blood: if a patient cannot legally have life-saving treatment forced upon her, it is entirely unreasonable to suppose that non-life-saving treatment should be forced on someone in Mr Blood's situation.

Clinicians' decisions in these situations should focus on the narrow, medical sense of best interests. But perhaps we can still take account of the broader understanding of best interests if we move away from the idea that it should be doctors who make these decisions. Perhaps someone who is connected to the patient should be the arbiter of these dilemmas.

The spouse's testimony

There is certainly something appealing about the possibility of allowing the spouse to make judgements regarding the patient's best interests. As suggested, the interpretation of best interests varies hugely from person to person. One individual might be adamant that he never wants to become a father; for another, parenthood might be the ultimate goal in life. Doctors cannot be expected to know these facts about their patients, but patients' partners (or parents, friends, or other relatives, perhaps) may be better equipped to do so.

This is indeed the claim which Diane Blood made with respect to her husband. She was adamant that she knew he would have wanted his sperm removed to enable her to have his child posthumously. This was not mere speculation, but certainty, according to Mrs Blood. The doctors were not pushing her to have this procedure performed; rather it was her own conviction that drove events. And—in the absence of documented evidence—who could be better qualified than the spouse to testify as to the patient's best interests in such a context?

From this perspective, Mrs Blood's case looks extremely convincing. Only she possessed knowledge of what her husband had wanted, and this was not merely based on vague inferences drawn from observations of her husband's character or his moral convictions. Mrs Blood said that she and her husband had previously discussed the question of posthumous births. Before Mr Blood's illness, she and her husband had read a newspaper article that described a case of posthumous insemination. The man in question had provided the sperm while still alive, and after his death his wife had been inseminated with the stored sperm.

Mr and Mrs Blood were moved by this story. It prompted them to discuss the issue and they agreed that, in the event of Mr Blood's becoming ill or dying, this was a procedure that they would approve.

Despite their foresight in respect of informing each other of these wishes, the Bloods failed to record this conversation, or its outcome. (In any case, it should be noted that the discussion referred to a fundamentally different situation from the one in which the Bloods found themselves, since the man in the article had given his sperm during his lifetime, although the insemination took place after his death.) In the absence of documented evidence, Mrs Blood's request that doctors should retrieve her husband's sperm was based solely on her assertion that this was what her husband had wanted. Mrs Blood felt that her word—her testimony—was sufficient to justify the procedure which was carried out on her husband's body, and that further proof or evidence that this was what he would have wanted should not be necessary.

The clinician who performed the operation presumably also believed that Mrs Blood's testimony was sufficient to justify going ahead with the sperm retrieval. The difficulty here is that, clearly, while fathering posthumous children might have been Mr Blood's genuine desire, there was no objective proof that this was the case. There was no documented evidence; Mr Blood's illness had struck too swiftly even to allow a verbal discussion on the subject. Perhaps Mrs Blood was indeed the person best placed to advise on Mr Blood's best interests in the broad sense discussed here, but she had no way of substantiating her belief regarding her husband's wishes. The clinicians who took the sperm simply took her at her word.

It is necessary to observe here that no-one wants to impugn Mrs Blood's honesty and integrity on a personal basis. As an individual, Mrs Blood made a convincing case, sufficiently so that the clinician who removed the sperm was moved to do so in the absence of formal consent from the patient himself. However, both from the perspective of Mrs Blood's case as an individual, and on a broader social level, there is a serious problem. Mrs Blood's convincing argument that these were her husband's wishes was clearly accompanied by an overwhelming desire to have her dead husband's children.

The fact that Mrs Blood herself had an interest to pursue meant that there was a possible conflict between this interest and her ability to give an accurate and unbiased account of her husband's interests. A conflict of interest is not something to be treated lightly in the context of surgical interventions. The best of judges acknowledge themselves to be incapable of impartiality in certain situations. Those who are entrusted with judgements which affect other people are excluded from making these decisions if they are deemed to have a conflict of interest. There is nothing shameful or immoral about being in this position, provided that the conflict is acknowledged.

In the case of Diane Blood, the danger of a conflict of interest was not addressed. Here the broader significance of this case begins to emerge. Because even if Mrs Blood was telling the truth, and even if doctors and lawyers were convinced of this, simply accepting her word on the matter has far-reaching implications. Since doctors were willing to remove sperm without consent in the case of Mr Blood, would they be prepared to do so in other cases, perhaps with less convincing spouses? If not, how could they justify their refusal?

After the instigation of a concerted media campaign, many people may have had a 'gut feeling' that Diane Blood was telling the truth when she declared that her husband would have consented to the procedure that allowed her to become the mother of his child. But gut feelings in themselves are simply not adequate in law to protect individuals from assault. Nor can they make for just laws, which—after all—have to apply to everyone. If a similar case arose, the doctor might intuitively feel that the spouse was lying and refuse the procedure. In the absence of specific protocols on consent, a reliance on gut feelings as a means of identifying which individuals' desires are met, and which are not, would rightly be regarded as arbitrary and inconsistent.

A person whose request had been denied on the strength of a gut feeling held by doctors might well claim that she had been unfairly discriminated against. Again, this seems to place too much power in the doctor as an arbiter of non-medical judgements. Moreover, people's opinions might well differ. Some individuals might trust a particular tearful spouse, while others might feel sceptical. The problem is that, compelling as the individual case may be, it is simply inescapable that a relaxation of the rules in order to accommodate a particularly appealing case creates a precedent which undermines the law.

Comatose and unconscious patients represent some of the most vulnerable members of society. It is true, although unfortunate, to say that where people are vulnerable, they may be looked upon by others as 'fair game' or as repositories of goods which can be obtained for the benefit of others. It is specifically for this reason that our laws are so stringently drawn up to protect these vulnerable individuals, and if the law is to be respected, it must not simply be waived every time that an appealing or emotive case arises.

If the law was allowed to be undermined in this way, so that written consent was no longer a requirement for the retrieval of tissues from the bodies of the comatose, dead, or dying, it is not hard to imagine that people's dying bodies might habitually be stripped of tissues and organs regardless of what their wishes might have been. Eggs and sperm are fast becoming recognized as being extremely valuable commodities both in fertility treatments and in research. Scarcity of these resources has resulted in the suggestion that financial incentives be offered for gamete donors, and that women should be allowed to 'pay' for certain medical procedures with their eggs.

Suppose these highly prized and valuable tissues could be removed from comatose or dead patients without prior consent. This would put patients at risk of being violated for the sole purpose of obtaining them. The prospect of women such as Mrs Blood (or indeed men in similar situations) wishing to have children by their dying spouses' gametes is simply one side of the coin. The possibility of obtaining commodities with a market value from comatose patients without their consent might in itself constitute a temptation for a spouse facing an impoverished future, for example. And even though the open selling of gametes is not yet permitted in the UK, it should be remembered that international markets in body parts are not unknown.[6]

In this respect, at any rate, it is clear that there is a distinction to be made between those interventions which might be undertaken, say, in order to fight an infection or to prolong the patient's life, and those which are undertaken for neither of these reasons, and at the instigation of a third party who also has an avowed stake in the outcome. In the latter case, the need for robust consent procedures would appear to be vital if vulnerable patients' tissues are not to become the property of those who may have something to gain from them. If comatose patients' bodies are to be invaded at the behest of people who may have an interest in the contents, we enter a realm where the possibilities begin to look truly gruesome. Why stop at gametes? Why not extract other tissues or organs . . . ?

Suppose Mrs Blood had been on the waiting list for a kidney transplant, and that Mr Blood was a tissue match, although not a donor card carrier. While he was comatose, Mrs Blood could have mentioned to doctors that he had wished to donate a kidney to her, but had never committed this to writing. In such a situation, her obvious interest in the outcome would surely skew the possibility of establishing whether this really was the patient's desire. Allowing this kind of organ harvesting on the say-so of the relative or partner would create a clear incentive for people to urge the cessation of life support, or the refusal of life-sustaining treatment for their relatives or partners.

Of course, organs are sometimes removed from the bodies of brain-dead individuals even if there is no written evidence that this was the wish of the dying patient, albeit under different legislation. It might well be argued that the position with respect to gametes and organs is inconsistent, but they are regarded as distinct issues in English law. Despite this, where a relative has an interest in the organ, it seems much more unlikely that organ removal from a dying person would be acceptable in the absence of a donor card.

One further point to make here is that the removal of gametes may result in the perpetuation of the patient's genetic line in a way which, for example, liver or kidney transplantation does not. The removal of gametes means that a person may become a parent after death. Typically, we regard ourselves as having an interest in controlling our reproductive destinies. This interest may cause us to view gametes differently from other tissues. Certainly, those who carry organ donor cards cannot thereby be assumed to be willing to donate gametes.

Some further implications

It is useful to think about how the Blood case might have looked if things had been the other way round. Suppose that Mrs Blood had contracted meningitis, and that her husband wanted to use her eggs to have a child? As we know, the collection of eggs requires surgical intervention: it is an intrusion on the body which might be regarded as an assault, unless mitigated either by the existence of written consent, or by the fact that it is medically indicated (ie necessary for the treatment of disease or in order to preserve the patient's life).

Should we be happy for this kind of intervention to take place in the absence of the appropriate consent? It seems unlikely, again partly because eggs (even more than sperm) are to some degree marketable commodities in today's world, both in reproductive therapies and for research. Even if the eggs were sought by a man desperate to have his partner's baby, the problem would remain that a man cannot simply obtain the eggs and do the rest himself. He would need the help of a surrogate mother. But if we allowed the removal of eggs, why not also allow the use of the uterus? In fact, pregnancies in comatose or dying women are not unknown.

Case 22: Susan Torres[7]

Susan Torres was a 26-year-old expectant mother, apparently in good health, and eagerly anticipating the birth of her second child, when tragedy struck. The family had just sat down for their evening meal together when Mrs Torres suffered a stroke, caused by a previously undetected brain tumour.

Mrs Torres' condition rapidly deteriorated, and doctors were unable to save her. Jason Torres, her husband, was informed that his wife's brain function had ceased. Only a life support machine was now keeping her alive. Mr Torres was asked whether the machine should be switched off, and his wife be allowed to pass away. However, Mr Torres made an unusual request: instead of switching off the life-support machine, he suggested to hospital staff that life support should be continued until the foetus his wife was carrying could survive outside its mother's body.

This proposal was accepted by the hospital staff, and Mrs Torres was kept on life-support as her unborn baby continued to develop inside her. At the time of her hospitalization, Mrs Torres had been fifteen weeks pregnant. A foetus has a chance of survival from around twenty-four weeks; however forty weeks is the duration of a normal pregnancy and it was hoped that Mrs Torres could be kept alive as long as possible to enable the child to have the best chance at survival.

Two months later, the baby was delivered by caesarean section, weighing under 2lbs, since it was determined that Mrs Torres' condition was deteriorating. Shortly afterwards, Mr Torres made the decision to have his wife disconnected from the life support machine. Initially, the baby—a girl, named Susan, after her mother—appeared to be doing well. However, she died five weeks later from complications associated with her premature birth.

The case outlined above made media headlines in 2005. It is of interest here not just because of the emotive interest it arouses, but because it highlights the fact that it is possible for unborn children to continue developing inside a comatose mother. In the case described above, the mother was already pregnant at the time that she became comatose. However, it is important to consider the possibilities which modern technology offers, and look at their potential applications.

If Mrs Blood had suffered meningitis and been irretrievably unconscious, perhaps Mr Blood would have been eager to keep open the possibility of having a child with her. Although she was not pregnant at the time she fell ill, he might mention that they had been trying for a baby. Mr Blood might cite a previous agreement made with his wife that if one of them should fall ill the other should pursue every possible avenue to have their spouse's child.

One way of achieving this would be to collect eggs from Mrs Blood, fertilize them with his sperm, and re-implant them in her uterus. Life support could then be continued for as long as necessary to ensure the maximum chance of survival for their offspring. Not only this, but

Mr Blood might well argue that in fact life support should be continued after the birth of the child in order to facilitate the birth of siblings. In effect, Mrs Blood's reproductive capacity could be prolonged despite her brain death, until the Blood family had reached the desired number of children.

This scenario would be regarded with horror by many people. However, it is technologically feasible, and is merely a logical extension of the idea that the bodies of the irrevocably unconscious can be utilized to further the needs and desires of other people.

This example should show us how important it is to retain robust consent procedures, and not to let these protective legal barriers be eroded by emotive individual cases.

There is another angle which could be considered here. As suggested, the removal of gametes without the patient's consent does not seem to be justified either ethically, in terms of the law, or in terms of our commonly-held assumptions and beliefs. However, at the time when Mr Blood's gametes were removed, the law had not been tested on this point. Ironically, one result of Mrs Blood's campaign was to publicize the fact that removing gametes in these circumstances was not regarded as being permissible. While 'pre-Blood' clinicians might have been unaware of this fact, 'post-Blood' clinicians could hardly claim ignorance. For this reason, it was felt that the Blood case represented a one-off situation which could not recur in the future.

The law has indeed been clarified, and it seems unlikely that future clinicians would want to risk facing litigation for removing gametes without the appropriate consent. It should be mentioned here, however, that while clinicians may be aware of the change in the law, the public are not necessarily so. Many people assumed that Mrs Blood's court victory over the HFEA meant that the removal of gametes without consent had been lawful. This was not the case. Mrs Blood's victory turned on the issue of whether or not the HFEA had the right to prevent her from exporting the sperm. The two cases in the High Court and the subsequent appeal concerned the issue of consent to the usage and storage of sperm as required in the HFE Act, as well as judicial review of the HFEA's use of its discretionary powers under the HFE Act to refuse to allow export of the sperm to Belgium.

Under section 24(4) of the HFE Act the HFEA has discretion to allow the import and export of gametes into and out of the country subject to conditions, and has the power to waive the requirements of sections 12 to 14 (concerning sale, records, welfare, and consent) on those occasions. The HFEA refused Mrs Blood's request to have the sperm exported to Belgium, largely on the ground that the request was being made in order to avoid UK law governing consent and the taking of the sperm.

If this were not the HFEA's policy, then serious breaches of the law could be overcome relatively easily. Allowing the export of gametes or embryos to another country where there are no relevant regulations, or where regulations are looser, would call into question the whole purpose of having regulation. However, this policy can conflict with European laws calling for freedom of movement in relation to provision of services, as indeed happened in the Blood case. But the HFEA did not consider that those provisions of the Treaty of Rome could possibly cancel out the need for strict maintenance of the UK regulatory system, with its benefits of health and identification. Individual European countries have the right to diverge from European provisions where there are strong national reasons relating to cultural and social matters best dealt with by the host nation, which seem to it to outweigh the benefit of the European provisions.

Nevertheless, the Court of Appeal found that the HFEA had not sufficiently weighed up the European principle of freedom to seek medical services when reaching its discretionary decision not to allow Mrs Blood to export her husband's sperm, and the Court decreed that the HFEA should take its decision afresh in the light of the judgment (that is to say, the HFEA's decision was not overturned: it was merely required to reconsider). The Court also held that if Mrs Blood were permitted to take the sperm abroad, the fact that it had been unlawfully obtained and stored should not be an impediment.

A significant feature in the Court of Appeal's deliberations was its belief that this peculiar situation was (and would always remain) unique. In response to the HFEA's fears that the granting of judicial review would mean that every widow in similar circumstances would seek to circumvent the law, the court suggested that no doctor would ever be likely to remove gametes in such circumstances again, as all would know that the laws of consent in the HFE Act could not be fulfilled and that storage and use would be illegal.

The spectre of hundreds, or even thousands, of people all desperate to remove sperm or eggs from dying relatives may have seemed unlikely or even ridiculous to those who were not involved in the day-to-day workings of the HFEA. However, it is both alarming and edifying to note that following the publicity associated with the Blood case, the HFEA was besieged with petitions and requests outlining tragic circumstances in which loved ones had died (mainly due to traffic accidents), and requesting the extraction of sperm and eggs despite the lack of written consent.

Not all of these claimants were the partners or spouses of the individual in question: in some cases the parents were the ones making the claim. One particular case involved the parents of a young man who had been left brain-dead following a motorcycle accident. The young man did not, in fact, have a wife or girlfriend, but his parents wished to obtain his sperm nevertheless, in the hope that at some stage in the future they might find a surrogate and egg donor willing to carry the child, thus retaining some hope for them of having a grandchild who was genetically related to their son.

The fact that so many people came forward shows the extent to which spouses, partners and other relatives have an interest and a stake in their loved ones' gametes. Yet it also demonstrates the danger that a conflict of interests may emerge: for some of these people, the urge to obtain gametes was sufficiently strong to impel them to support their case with documentary evidence which was quite clearly spurious. These claims were refused.

Although this revealed the scale of possible interest in obtaining posthumous gametes, to some extent Mrs Blood's case *was* unique. As the Court of Appeal had observed, in that case, the gametes had already been obtained, and even though the process by which they had been obtained might be regarded as unethical and unlawful, there remained a separate question about what should be done with the unlawfully obtained sperm.

Part of Mrs Blood's argument was that, even though the sperm had been procured in a way which breached the consent requirement, it might as well be used now that it had been obtained. What would it serve to prevent its use? The damage had already be done; if there was anything to be salvaged from this situation, it would surely be to accept the past, learn from the mistakes that had been made, but not unfairly to penalize a woman who had acted in good faith throughout. No-one had accused Mrs Blood of any wrongdoing; nor could she have been expected to have an in-depth knowledge of the law in this area. She had trusted the doctors to make these judgements, and they had made serious errors. But now that the sperm was at a clinic there seemed little point in preventing her from using it.

However, while this may have seemed to be the common sense view, the HFEA took a different line. Posthumous use of gametes is specifically prohibited by the HFE Act, in the absence of written consent. Indeed, gametes may not lawfully be stored, or used in treatment at all, unless there is specific written consent. There is only one exception to this rule, and that is where a man and woman are receiving treatment together. The reason for this exception is to allow for circumstances where a man and woman are both present, and the man is to produce sperm for immediate insemination, ie when storage of the sperm is not at issue. In these circumstances, the willingness of the man to undergo the procedure is deemed to imply his consent for the use of the sperm in that way, and at that time. Clearly, Mr and Mrs Blood could not have been said to have been receiving treatment together, since Mr Blood was already dead. Therefore the necessity of consent for the storage and use of his sperm could not be overcome on these grounds.

The seriousness of the breach of consent requirements seemed to militate against simply allowing the unlawfully obtained sperm to be used in treatment. One of the main points of the HFEA's existence is to ensure that fertility treatments in the UK are strictly regulated. This means that the entire process must come under strict scrutiny. There is little purpose in attempting to ensure that clinics are operating appropriately if the very gametes being used have been obtained under dubious circumstances. To allow Mrs Blood to access the sperm would seem to make a mockery of the legal and ethical framework which was designed to maintain public confidence in the UK's fertility services.

While an individual may feel that the law is unjust when it thwarts his or her personal wishes, laws are not framed for individuals, but for society at large. This means that transgressions cannot necessarily be excused, nor a blind eye turned to particularly emotive cases. The law must behave with impartiality in all its dealings. In Mrs Blood's case, sperm had been illegally obtained from a patient, and was continuing to be illegally stored. To facilitate treatment with this sperm would be to bend the law and—in essence—to suggest that while consent requirements are important, they can be circumvented, or waived for particular cases. If regulation of fertility treatments is desirable at all (and of course, those who think it is not desirable will not agree here) it is imperative that the HFEA does not turn a blind eye to treatment with gametes which have been obtained unlawfully.

Sperm and eggs, as discussed earlier, are marketable commodities. Obtaining them can involve the violation of people's bodies. Using them can entail that a person becomes a parent, that his or her genetic line gets transmitted. These are fundamentally important aspects of our lives. The HFEA needs to be confident that the gametes used in the treatments which it regulates have been obtained according to rigorous ethical and legal protocols. If these protocols are broken, the power of the HFEA to regulate effectively is called into question. Clearly, the HFEA cannot prevent the infringement of a patient's autonomy, or the violation of his body after this has happened. However, it can and should do everything in its power to ensure that gametes obtained in these ways are not used in treatments under the HFEA's jurisdiction.

Suppose, as is not beyond the realm of possibility, eggs being stored at clinics in the UK were found to have been removed illegally and without consent from women in disadvantaged circumstances. Suppose, even, that eggs were being taken from the bodies of unconscious women without their consent. Again, the crime would already have been committed, yet surely we would not be justified in thinking, 'oh well, eggs are desperately needed, we may as well use them now that we have got them'. To do so would in effect constitute an incentive for those who are tempted to side-step the legal and ethical protocols around gamete procurement.

The need to maintain rigorous and even-handed controls over the controversial issues that arise from fertility treatments means that occasionally a deserving and well-meaning individual cannot have what he or she wants, because this desire flouts the safeguards which are erected to protect the vulnerable. We cannot claim the protection of the law except when it turns out to be disadvantageous to us, unless it is arbitrary, discriminatory, or unjust. In the case of Diane Blood, it was none of these.

Moral responsibility and the media

The field of reproductive technologies is so swiftly-changing that many of the challenges which present themselves to the HFEA are in some way unique, or would create a precedent. This means that conflicts between individual desires and the public interest often come to the fore. As discussed, there are many good reasons why Diane Blood should not have been permitted to use gametes which had been obtained unlawfully. However, the HFEA was outmanoeuvred on this issue, and it is necessary to point out here the role played by the media.[8]

Sympathetic to Mrs Blood's case, reporters were unlikely to take the dreary, mundane view that the public interest needed to be served by preventing Mr Blood's sperm from being used. Having taken a supportive position towards Mrs Blood, virtually all the British newspapers ignored the way in which Mr Blood's sperm was obtained. In some cases, reporters seemed to imply that Mr Blood had somehow given his consent to the procedure.[9] The technique for extracting sperm was a central feature of this case, and might well affect people's perceptions about the acceptability of waiving the law in these situations. Keeping silent on this issue allowed the news reports to skim over this issue and focus on other, more palatable, aspects of the case.

The media, of course, was bound to be interested in an emotive situation such as this. But is it appropriate that the media should play an active part in influencing the outcome in such a case? The degree to which the media involved itself in influencing the course of events is revealed by the fact that a fund established to meet Mrs Blood's legal expenses was heavily promoted by the *Daily Mirror*.[10] Despite popular opinion, the HFEA's funds to fight legal battles are severely constrained, and this means that some cases cannot be pursued however damning the evidence, and however publicly important the issue. In these circumstances, it is the stark truth that access to funds may ultimately be the factor that spells failure or success. It is also worth noting here that, in any case, costs were eventually awarded in Diane Blood's favour.

The issues raised by the Blood case were felt to be a matter of public interest by the HFEA. But for the media, it was far more exciting to champion a widow in distress, and generate a surge of warm-hearted concern, especially when campaigning for Mrs Blood enabled just the kind of approach and language which makes readers feel good and sells newspapers. Consider the alternative headlines: 'Consent remains a paramount concern'; 'Woman does not have baby'. The journalists wanted Mrs Blood to succeed: they would certainly get more mileage from the case if this happened and a child was born.

Of course, it is not the role of the media to act solely with the public interest at heart. Individual sob stories are what sell newspapers, and we, as the people who read those newspapers, are all implicated in this. However, in this case the media's take on an individual case came into conflict with the regulatory safeguards designed to protect patients from assault. In the event of such a conflict, it may not be desirable that the media should prevail, nor that it should sway the outcome of our established legal processes.

Had money and media pressures been more relaxed, the HFEA might well have been able to establish that there was a public interest to be protected in the Blood case which justified an infringement of the European right of provision of services. If we are to resist the levelling down of our national legal standards, these considerations in our law need to be expressed as protected principles. They are indeed the broad principles by which the HFEA operates: human dignity and autonomy expressed in the right of informed consent which may be withdrawn at any time; the welfare of the potential child; and the maximum reasonable standard of safety of treatment for the persons involved.

The eventual decision on the Diane Blood case was a capitulation in the face of heavy pressure. The HFEA had been proved right with respect to questions relating to domestic law. However, it was forcefully suggested that its jurisdiction did not necessarily justify a refusal to allow the export of gametes. Under European law, this kind of import and export cannot legitimately be curtailed except under exceptional circumstances. The HFEA was invited to reconsider its refusal to export. In effect, however, it had run out of resources to continue the fight. There was little doubt that in the face of such huge public and media pressure, a refusal would have been impossible.

Elsewhere in this book it has been suggested that public involvement with policy and decision making is highly important. However, it is vital to distinguish between emotive outpourings fanned by heavily biased information supplied by the media, and public consultations undertaken on the basis of full and independent information. Public consultations usually focus around abstract or policy issues. News stories such as that of Diane Blood necessarily revolve around the individual, and as an individual of course it was desirable that she should achieve what she wanted. The media involvement in this case forced an outcome which was based on skewed facts and distorted public perceptions. The fundamental legal questions and implications were swept aside. If judgements of this kind become the norm we may all ultimately be the losers.

After emerging victorious from the legal battle on the export of her husband's gametes, Diane Blood received fertility treatment in a Belgian clinic, and gave birth to a son. A further course of treatment a few years later was also successful, and Mrs Blood is now the mother of two boys.

The issues related to her case were debated in Parliament and subsequently a review of the law of written consent in medical treatment was undertaken.[11] Its findings were reported at the end of 1998, just before Mrs Blood gave birth to her first child some three years after her husband's death. The report concluded that the law was correct to insist on written consent, and should not be changed. The HFEA was recommended expressly not to allow export of sperm to overcome breaches of British law.

CHAPTER 6
Human Rights and Reproduction

Human rights legislation has only relatively recently begun to have an effect on British law. The European Convention on Human Rights was enacted in the UK by the Human Rights Act 1998, which was implemented in 2000. Its statements of the right to respect for private and family life and the right to marry and found a family have had a significant impact on reproductive medical law. The invocation of human rights in a case such as that of Kirk Dickson, described below, demonstrates the challenges that these new legislative frameworks raise.

Prisoners and the right to reproduce

Case 23: Kirk Dickson[1]

In 1995 Kirk Dickson was convicted of murder and was sentenced to life imprisonment. While serving his sentence, he began corresponding with Lorraine Earle, who was serving time for benefit fraud. The couple formed a relationship via the prison pen-pal scheme, and were married after Earle's release in 2000. Dickson himself was not due for release until 2009.

Shortly after their marriage, the couple began to explore ways in which they might achieve their desire of starting a family. Since Mr Dickson's new wife was already over 40 years old there was clearly no time to lose. Mr Dickson was not entitled to conjugal visits: the only possibility seemed to be artificial insemination, as Lorraine would be over 50 if she waited for Mr Dickson's release, and would have little hope of becoming pregnant naturally.

Accordingly, the couple sought permission for Mr Dickson to provide a sperm sample with which his wife could be inseminated. However, the application was refused. After further setbacks from the High Court and the Court of Appeal, the couple decided to take their case to the European Court of Human Rights. It was argued that the previous rulings had violated Mr Dickson's right to family and private life, as well as his right to marry and found a family.

The couple's initial case was turned down by the European Court of Human Rights in Strasbourg. However, not to be deterred, they persevered with their claim, taking it to the Grand Chamber. At the time of writing, a final decision has not yet been reached.

Mrs Dickson already has three children; Mr Dickson has none.

For many people it might seem absurd that prisoners serving life sentences should think they have a right to access assisted conception. However, the case above is not an isolated incident,[2] and criticisms have been made of the fact that British prisoners do not have either conjugal rights or access to fertility treatments.

The question here is whether any of the rights listed in the Act support a specific right to have children. And if so, are there some circumstances in which it is reasonable to override this right? Perhaps still more importantly, it should be established whether a right to reproduce implies that there is also a right to receive assistance with reproduction. If so, this could prove extremely expensive.

Clearly, there may be some occasions when people's human rights are forfeited, perhaps because they have embarked on actions which infringed the rights of others. The court in the case described above did not find that Dickson had a specific right to reproduce. Prisoners do not necessarily lose all their rights, but the right to have children, at least by natural means, is contingent on the very things which by definition prisoners have lost: privacy and freedom.

What about Dickson's wife? Does she have a right to reproduce? In this case, Mrs Dickson herself had committed a crime, and indeed this was how the couple became acquainted. However she had served her sentence, and it would seem extremely harsh to say that she should be subject to restrictions on her reproductive freedom simply because her husband happened to be in jail.

Yet a moment's thought reveals that those connected with convicted criminals frequently suffer as a result of the incarceration of that person. Family income may be adversely affected. Spouses may struggle to care for their families unaided by the imprisoned partner and children may suffer by being deprived of one of their parents. Employers—or employees—may be left in the lurch. Even though these people may have done nothing to merit their own rights being infringed, it is generally regarded that they are simply unfortunate to be connected with the prisoner in this way. There is thought to be a greater social interest at stake: that of enforcing the law.

Another point here is that the reproductive rights of Dickson's wife were not necessarily infringed per se, as she could have a child with another person if she so desired—just not with Mr Dickson.

The case of Mr Dickson highlights the difficulty of establishing exactly what a right to reproduce might entail. Dickson cited his 'right to marry and found a family' in his argument. He interpreted the right to 'found a family' as meaning that the State had no grounds to prevent him taking such steps as were necessary to have children. However, in the context of his imprisonment, non-intervention would not be sufficient to meet his needs, and the State would have had to take active steps to facilitate his reproductive endeavour.

As we have seen, the UK and European courts' decisions implied that the State has no obligation to facilitate prisoners' reproductive enterprises. But some countries take a very different view.

Case 24: Yigal Amir

In 1995, the prime minister of Israel, Yitzhak Rabin, was assassinated. Right wing radical Yigal Amir was arrested for the crime, tried, and sentenced to life imprisonment. While serving his sentence, Amir began corresponding with Larissa Trimbobler, who sympathized with his extremist views. In 2004, Amir and Trimbobler were married by proxy.

The couple were eager to start a family, but conjugal visits had been denied to them, as they were deemed to be a security risk. Amir requested that he be allowed to use artificial insemination in order to impregnate his wife. While the petition was being considered, Amir was found to have attempted to smuggle a test-tube filled with his sperm to his wife. The attempt was thwarted.

In late October 2006, after a dramatic turn-around in policy, Amir was permitted a 10-hour conjugal visit with his wife. As they had hoped, she became pregnant.

Not all States adopt the same position with reference to prisoners' rights to reproduce or to have sex with their partners, and in Israel, prisoners are allowed to request conjugal visits. In initially debarring Amir from this opportunity, the Israeli authorities may have been responding to the particular repugnance felt for his crime. However, when faced with Amir's demand for positive intervention in the form of fertility treatment, conjugal visits may have seemed to be the lesser of the two evils.

In contrast, the European court's ruling on Mr Dickson's case turned on the idea that where a person has forfeited the right of freedom, other rights which are based on this right will also be forfeited. In Israel, the reluctance to encroach on prisoners' reproductive lives seems to weigh against this interpretation. Yet as shown, this can lead to some uncomfortable conclusions. These national differences highlight the varying cultural importance of reproduction.

However, this does not directly answer the fundamental problem of reproductive rights. Is there a right to have a child, as such? Certainly, no such right is specified in the 1998 Act. However it is argued that some of those rights which are protected in the Act can be interpreted so as to encompass a right to reproduce.

Here, it is useful to establish which rights are relevant in the context of assisted reproduction.

The rights in question

Articles from the Human Rights Act 1998

Article 3

Prohibition of torture

No-one shall be subjected to torture or to inhuman or degrading treatment or punishment.

Article 5

Right to liberty and security

1. Everyone has the right to liberty and security of person.

Article 8

Right to respect for private and family life

1. Everyone has the right to respect for his private and family life, his home and his correspondence.
2. There shall be no interference by a public authority with the exercise of this right except such as is in accordance with the law and is necessary in a democratic

society in the interests of national security, public safety or the economic well-being of the country, for the prevention of disorder or crime, for the protection of health or morals, or for the protection of the rights and freedoms of others.

Article 12
Right to marry

Men and women of marriageable age have the right to marry and to found a family, according to the national laws governing the exercise of this right.

Article 14
Prohibition of discrimination

The enjoyment of the rights and freedoms set forth in this Convention shall be secured without discrimination on any ground such as sex, race, colour, language, religion, political or other opinion, national or social origin, association with a national minority, property, birth or other status.

Article 3 prohibits torture or degrading treatment or punishment. Could this encompass a right to reproduce? Perhaps being forcibly separated from a willing sexual partner might constitute degrading treatment of this sort. However, it is not clear that simply failing to provide access to such a partner would fall within the scope of this right.

Despite this, there are some disturbing considerations with respect to the case of Amir. Conjugal rights, as mentioned, were allowed in this case, and indeed, are often permitted for Israeli prisoners. However, Amir's conjugal visit took place in a context of heavy surveillance (on account of the security concerns). One may imagine the security personnel keeping a close eye on the proceedings, and it is hard to imagine this could have been welcomed by either of the protagonists. Is there something degrading about such treatment? Perhaps so.

But this does not necessarily argue in favour of a right to reproduce. If the treatment had been deemed inhuman or degrading, this could have been remedied by either by providing greater privacy, or by refusing conjugal visits altogether. It does not seem to relate specifically to issues concerning reproduction.

Article 5, the right to liberty and security, seems scarcely to justify a specific right to reproduce in the context of prisoners. For non-prisoners also, it would seem dubious to infer that this right implies a corresponding right to reproduce. This is primarily because however much liberty and security one has, the ability to reproduce depends on other factors, and requires the willing co-operation of at least one other person. This issue is explored in greater depth later in the chapter.

Articles 8 and 12, the right to private and family life, and the right to marry and found a family, are the two most often suggested as the basis for a right to reproduce. But the right to private and family life is by no means absolute. It may be overridden for a number of reasons given within the Act itself. Most obviously, it may be overridden on the basis of the law, or to prevent crime. So if a person is legally imprisoned, any rights to privacy and family life may be trumped. This is not to say that there is a clear case to be made on these grounds here, but simply that even from within a rights perspective, claims can be rejected.

There may be something unsatisfactory about the degree to which these rights are conditional. For example, the right to privacy and family life can be overridden to prevent disorder, or to protect public morals. Yet since there is no indication of what these rather vague concepts mean, it seems possible that some very restrictive States could interpret this in such a way as to leave their citizens no remnant of their right whatsoever.

The right to marry is an interesting anomaly in the context of prisoners. British prisoners are not prevented from marrying, despite the fact that they cannot receive conjugal visits. This may be connected with the issue raised above. Making provision for a prisoner's sexual requirements creates an uncomfortable clash between notions of privacy and the need for security. Allowing a prisoner to take part in a formal legal or religious ceremony may seem less problematic. Again, though, in the wider context, the right to marriage cannot necessarily be seen as implying a right to reproduce, for reasons that will be discussed later in this chapter.

Article 14, the right to be free from discrimination, might be thought to support a prisoner's claim to be able to reproduce. After all, non-prisoners can reproduce, so it is discriminatory to prevent prisoners from doing so. However it is obvious that prisoners *are* discriminated against: to deny this would be nonsense. If we disapprove of all forms of discrimination, we should not be willing to countenance imprisoning people at all.

The discrimination exercised against a prisoner is, in any case, not based on any of the criteria set out in the Act (eg race, religion, etc). Once again, it simply follows from the legal status of the prisoner. Thus the fact that other members of the public are able to have sex, reproduce, and access fertility treatments does not per se mean that it is unreasonably discriminatory to prevent prisoners from doing the same things.

If certain categories of prisoner were allowed access to fertility treatments, the argument related to discrimination would have more force.

If Yigal Amir had been arguing his case from a rights perspective, this would have carried considerable weight, since some prisoners were allowed conjugal visits, and some were not. In the UK, the possibility of claims arising from patchy or inconsistent provision of conjugal visits or access to fertility treatment may be one reason for maintaining a universal restriction on such activities throughout the prison system.

In the Dickson case, as outlined above, restrictions on the prisoner's ability to procreate were a corollary of his having forfeited some of his rights as a punishment for his crime. Thus, simply failing to provide him with the means of procreating was not regarded as being an unacceptable infringement of his rights. A similar approach was taken in the case of Gavin Mellor, a UK prisoner whose case was also rejected both in the UK and by the European Court of Human Rights.

Prisoners forfeit their rights to privacy and liberty and it seems reasonable that they should forfeit other rights which are contingent on these. This means they cannot reproduce 'naturally'. But is it reasonable to infer from this that they should also be debarred from fertility treatment? The provision of conjugal visits may be so constrained by security concerns as to make them 'inhuman or degrading' because the privacy appropriate to sexual activity cannot be afforded. Alternatively, it might simply be too demanding in terms of prison resources. In contrast, allowing prisoners access to fertility treatments would not be subject to these difficulties. Indeed, denying access to medical treatment could be argued to be a contravention of Article 3, and it has been suggested that this Article could potentially form the basis of a claim to State-assisted fertility treatment.[3]

There are two points to be made. Firstly, prisoners' access to healthcare is permitted on the basis of medical need. It is not immediately clear that a fertile prisoner who has forfeited his liberty has a medical need for fertility treatment. The concept of medical need itself is a vexed one. Even when clinical infertility is diagnosed, it does not necessarily follow that the patient has a need for treatment, or indeed, a right to it. The second issue is that in popular terminology, reproduction is often regarded as a 'privilege' rather than a right. In this sense, perhaps it is conceptually separate from conventional medical treatment which may be necessary to relieve pain or save a patient's life.

Finally, the question remains whether it is appropriate to bring a child into the world in these circumstances. The British courts take this consideration into account in such cases. For many, enquiring too deeply into the suitability or otherwise of prospective parents is regarded as officious and intrusive. Yet the unquestioning provision of reproductive treatment might seem equally inappropriate.

In conclusion, prisoners do not currently have a default right to reproduce, at least in the UK. This stance can be justified with reference to provisions within the HFE Act itself, but the scope for interpretation means that other countries might well arrive at different conclusions.

Meanwhile, the question remains: if prisoners do not have a right to reproduce, does the general public have such a right?

The extreme view: the right to reproductive choice

Many of those who argue in favour of reproductive rights suggest that if an individual has a desire, the onus is on society to prove why that desire should not be met. Commentators such as John Harris, John Robertson, and others have argued that there is a right to reproduce, and that this right should encompass freedom of choice: the exercise of reproductive autonomy. For example, John Harris says, with regard to cloning:

> the significant ethical issue here is whether it would be morally defensible, by outlawing the creation of clones ... to deny a woman the chance to have the child she desperately seeks.[4]

From this viewpoint, people may have the right to choose what kind of offspring they have, as well as simply whether and when to have them. Attempts to interfere with these choices are considered to be unjustifiable infringements of people's rights. Some commentators are convinced that this right applies just as much to those who are infertile, and therefore need technological assistance, as to those who are able to conceive naturally. It is suggested that to prevent infertile couples from having children by means of reproductive technology is discriminatory unless there would also be intervention in the choices of fertile couples in similar circumstances.

This kind of argument has potentially far-reaching conclusions. For example, according to those who adopt this view, if a fertile couple with previous convictions for child abuse would not be prevented from conceiving naturally (for example, by forced sterilization or abortion), it could be unjustly discriminatory to prevent an infertile couple—also with convictions for child abuse—from using reproductive technology to have a child.

Alternative versions of reproductive rights arguments have been made by several feminist writers. Mary Anne Warren argues for the special nature of reproductive rights with relation to women. She suggests that even if there are aspects of reproductive choice which could carry negative social implications, 'reproductive freedoms are crucial to women's other basic moral rights, and it is wise to resist incursions upon these freedoms, unless the arguments against them are extremely strong'.[5] For Warren, reproductive rights include not only the choice of whether and when to reproduce, but also the selection of a child's attributes. Failures by the State to enable women to achieve the kind of children they desire are perceived as a sinister and unwarranted infringement of these rights.

This approach would constitute the basis of a strong argument for deregulation, but for many people this seems too extreme. Warren's assumptions provide no basis for discriminating between what is reasonable and what is not. For example, if a woman wants a child which possesses luminous jellyfish genes, should she be entitled to receive assistance with this desire as part of her reproductive rights? According to Warren's views, it would not seem obvious that this would be unreasonable, but intuitively this would seem absurd.

These arguments in favour of reproductive choice may seem appealing or ridiculous depending on one's inclinations. But before settling down to choose the attributes of our offspring, perhaps we should try to establish more satisfactorily whether a right to reproduce can in fact be drawn from any of the Articles listed above.

The right to privacy and family life: does it entail a right to reproduce?

Article 8 stipulates that family life should be respected. This has been interpreted to mean that individuals should be free to form families as they choose. But do we really know what constitutes a family, or family life?

Case 25: X, Y, and Z

In 1995 the case of *X, Y, and Z v UK*[6] appeared before the European Court of Human Rights. X, who was born a woman, had undergone surgical and hormonal gender reassignment and now lived life as a man. He had formed a stable and long term relationship with Y, a woman. Together, they wished to start a family.

Despite initial opposition, they eventually succeeded in being referred for treatment with donated sperm. Y was impregnated with the sperm, and gave birth to a child, Z. X could not be registered as the legal father of the child, since he was not biologically male.

Subsequently, X applied for, and was offered a job overseas. The conditions included free accommodation for dependants. However, because X could not be registered as the father of Z, the child did not count as a dependant.

In 1993, X, Y, and Z made a claim that their right to respect for family life had been violated by the failure to register X as Z's father.

This case shows the complexities that can arise when legal and social concepts of family life differ. The position of X here is interesting when compared to that of the husband of any woman who uses artificial insemination with donated sperm. In these circumstances the husband is regarded as the legal father of the offspring. In this case, X was excluded not because he had no biological relationship with the child, but simply because the UK courts regarded gender as being fixed at birth.[7]

The couple in the case described above had difficulties in accessing fertility treatment long before the issue related to legal fatherhood arose. They were eventually successful in persuading the decision-makers that they were indeed a couple, and would make suitable parents. It is interesting to consider whether they might have had a case if they had been denied treatment altogether. (This kind of issue and the question of what constitutes a family is considered more fully in Chapter 7.)

Respect for family life may feasibly entail that people's parental roles are recognized and acknowledged by the State, especially when a failure to do so results in loss of earnings or privileges, as in the case above. But it is not so clear that it would incorporate a right to have children. Rather, it seems to centre on relationships around children who already exist.

However, it has been argued that meeting the desire of any person for a child (by assisted reproductive technology or by other means) might be entailed by the right to privacy, also enshrined in Article 8.

> If the right to privacy means anything, it is the right of the individual, married or single, to be free of unwarranted governmental intrusion into matters so fundamentally affecting a person as the decision whether to bear or beget a child.[8]

Article 8 is certainly understood to make provision for the inclusion of sexual activities under the heading of 'private life'. But can the right to privacy really encompass a right to have a child? Although as a rule we may not intrude on people's privacy by forcibly preventing them from reproducing, the State can and does remove children from abusive families. Simply having had a child does not give the parent the right to bring it up. In cases where parents have harmed previous offspring and the mother has another child, the new baby may be removed *in advance* of any harm being perpetrated, simply on the grounds that remaining with the birth parent is too risky.

The apparently unlimited freedoms of the fertile are not as boundless as one might suppose. The reason that intervention does not take place before the mother gets pregnant is not that she is regarded as having a right to reproduce as such, but simply that in the relative hierarchy of moral concerns, interposing a physical impediment to people's sexual and reproductive activities is regarded as unjustifiably intrusive.

Seemingly, then, the right to privacy may encompass the right not to be prevented from conceiving naturally. But it is not clear what, if anything, follows from this.

In addition, while reproductive or sexual activity generally might be protected by the right to a private life, this is not an absolute protection based on the idea that sex or reproduction is intrinsically private. The performance of sexual acts in public, for example, is punishable by law. Attempts to reproduce through engaging in sexual activity—in private—may be justly regarded as beyond State intervention or control. However, once people step outside their homes and attempt to achieve the same goal in a public context, it becomes a matter of social, and possibly State, interest, and therefore loses the privileged private status which it previously had.

This suggests that those who seek fertility treatments are not protected by the right to privacy from State restrictions or interference. Infertility treatment is hardly an intimate or private activity. By its very definition it has to be performed in a clinic or hospital involving a number of doctors and other medical personnel. In the UK, as well as other European countries, the enterprise of embarking on fertility treatment is bound to involve regulation by public authorities, including form filling, the collection of data, reporting, and monitoring.

Positive and negative rights

The kind of reproductive freedom discussed so far could be construed as a negative right: a right not to be physically coerced in one's reproductive choices. In contrast, those needing treatment require positive intervention. It might be thought that this difference justifies a different level of intervention or State interest in assisted reproduction. But this distinction can be attacked:

> It may be argued that there is no real distinction between negative and positive rights. For if means are available and the state denies access to them that may be construed as an interference with freedom just as surely as compulsory sterilisation.[9]

Is it disingenuous to argue that merely failing to provide fertility treatment is somehow less offensive or unjust than physically preventing someone from reproducing, for example by subjecting that person to a forced abortion? It is fairly obvious that most people would not regard the two alternatives as being equally unwelcome. Although failure to provide fertility treatments might seem unfair, annoying, or deeply distressing, this would surely be an outrage of an entirely different order from the experience of being forced to undergo an abortion or sterilization.

It is vital that these two questions are not conflated. Women have campaigned long and hard to be free from physical coercion in their sexual and reproductive choices. It is only relatively recently that the law acknowledged the crime of rape where the victim is married to the perpetrator. Likewise, a case of a woman being forced to undergo a caesarean section against her will in the UK occurred as recently as 1998.[10]

Thus, there is a clear distinction to be made between positive and negative rights, at least in this respect. While a positive right to assisted conception may be desirable, its lack does not entail a violation of an individual's bodily integrity, and it is therefore to be regarded as a separate moral issue.

Rights and duties

Another facet of the distinction between positive and negative rights is the idea that they may have corresponding duties. As Margaret Brazier points out 'generally a right to x requires that there be a correlative duty on the part of some other person to supply x'.[11] If reproduction is a right, then it too must have corresponding duties. However, this may seem counter-intuitive. Who might be responsible for these duties? If a person decides that he or she wants to have children, does someone else have a duty to assist in whatever way possible?

Identifying those to whom this supposed duty falls could be extremely tricky. Perhaps the most obvious answer is that a person's partner is required to shoulder this duty. But insisting that a spouse has reproductive duties towards her husband seems unpalatable. This was not always the case. For example, it used to be impossible for a husband to be found guilty of raping his wife. The law has been changed on this point only relatively recently, and wives are no longer regarded as having a duty to fulfil the reproductive rights of their husbands. In the UK, women may choose to have an abortion or to use contraceptives, as it suits them or in negotiation with their partners. Conversely, a wife may be granted a divorce or annulment of marriage if the husband refuses to have sexual intercourse with her, but she cannot demand fertile intercourse as her right.

If the reproductive duties that correspond to reproductive rights are not owed by spouses to their partners, who *can* be said to owe them? Perhaps the State, or society as a whole. If this were the case, it might be assumed that there is a public duty to supply fertility treatments for those who may demand them. But this would be a very expensive undertaking. Nor is it obvious that such a duty can legitimately be imposed on the public. At the very least, if we have duties towards each other to provide certain goods, it is not clear that fertility treatments should be at the top of the list. Provision for basic food, housing, and healthcare needs might seem more important (and even in these respects, the public's willingness to provide for the needs of other human beings is extremely limited).

In some ways, it seems that the right to reproduce is a misnomer since it implies an individual power over something which is necessarily and intrinsically a process involving at least two people (and many more in the case of reproductive technologies). Reproduction is a feat that simply cannot be achieved on one's own, whether genetically or technically. Normally, of course, it is necessary to find a willing partner in order to be able to reproduce.

It might be tempting to think that cloning could offer a way of reproducing without the need to involve others, but in fact cloning would, if anything, require *more* willing collaborators than 'normal' reproduction. Cloning would involve a cell donor, an egg donor, and possibly a surrogate mother, as well as a team of scientists.

The right to marry

It is interesting here to compare the supposed right to reproduce with the right to marry set out in Article 12. Even the most enthusiastic of human rights supporters do not interpret this right to entail that the State, or any other person, has a duty to provide an individual with a spouse. Why not? Because a spouse is a human being rather than a commodity. A sexual or marital partner cannot simply be supplied on demand. Clearly, then, the proper interpretation of the right to marry is that it is a right *not to be prevented from doing so*. Crucially, this translates into a duty which we can all understand. While none of us—including the State—can undertake to provide spouses for other people, we *can* undertake to fulfil our duty of not interfering with those who have found potential spouses and wish to marry them.

Similarly, as we have established, fertile people do not have the right to have a baby naturally, since they cannot demand of any other individual the duties which would correspond to this right, such as the provision of a partner, or the obligation to have sex. But if fertile people do not have the right to have a baby, how can it possibly be the case that infertile people should have this right? This seems quite clearly absurd. Surely either everyone has the right to have a baby, or no-one does.

Article 12 of the Human Rights Act 1998, historically analysed, expresses our current distaste for the kind of enforced proscriptions of marriage in Nazi Germany and in South Africa under the apartheid era. A sensible interpretation of the Article is that—as far as possible—the State should adopt a non-interventionist policy in relation to its citizens' marital choices.

In all, then, a right to have a baby seems plausible only as part of a general right not to be prevented from marrying a willing and eligible partner, or from having sex. The upshot of these rights will be that there are very few circumstances in which fertile people can be legitimately prevented from reproducing. However, it is not reproduction per se that is protected.

Rights and regulation

If we accept the existence of specific reproductive rights, as opposed to the negative rights described above, these will come at a cost in terms of resources. A duty to meet these reproductive rights would surely fall upon the State (as there are no individuals who can feasibly be held responsible for the reproductive requirements of others). Therefore, in a democracy, it is the members of society who must acknowledge or repudiate the right for everyone to have access to ART.

In its approach to regulation in the fast-moving world of reproductive technologies, the HFEA has taken the line that there is no 'right to a baby' as such. The modern age is sometimes described as one in which everyone has rights and every desire must be met. The HFE Act, however, enforces 'the ethics of the nation' rather than simply regarding the individuals' desires as an imperative.

In common with other authorities, the HFEA is now bound by section 6 of the Human Rights Act 1998. This requires public authorities to act in compliance with the Convention in all aspects of their work, unless there is a specific statutory obligation to the contrary. One of the upshots of this is that where existing statutes conflict, they may be resolved with reference to human rights principles. Since a failure to comply with human rights obligations is a ground of legal attack, bodies such as the HFEA must ensure that their procedures do not run counter to the rights legislation. However, given the latitude of interpretation that the rights legislation allows, this is not always as simple as it sounds.

When individuals feel that the HFEA is denying them something which they can claim as a 'right', this can result in costly court cases. Conflicts often arise between the perceived rights of patients or clinicians, and broader social concerns. Assisted reproduction is continually breaking new ground and challenging old assumptions. This may mean that statutes governing the possibilities in this field become quickly outmoded in the light of new developments, and in the context of new applications of existing techniques.

Statutory uncertainties and ambiguities leave loopholes for those who are desperately seeking to interpret the law in favour of their chances of having a child. There are also plenty of lawyers eager to make their reputations in fertility cases, with their popular settings and outcomes.

It is an open question whether the human rights legislation could persuade a judge that every woman, no matter how old or young, no matter what her condition and status, has the right to attempt IVF, whether privately or through the NHS.

It is even more open to question whether the legislation should persuade a court that every woman has a right to treatment designed to maximize her chances of becoming pregnant. Certainly this is not always the case, as demonstrated in the Taranissi judgment below, a decision which encompassed consideration of the human rights legislation.

One cannot necessarily blame would-be parents for seeking to identify every possible opportunity which might give them a chance to have a baby. In this context it is easy to see the appeal of rights rhetoric, which gives the impression that rights are discoverable; as though one can examine the nature of human beings and thereby divine what kinds of things they should have. However, on the model suggested here, rights are far weaker, and are established through social agreement rather than discovered as pre-existent entities.

The individual and society

A society may decide that it wants every individual to have the right to be free from the danger of starvation, or to receive free education. But each decision of this nature carries an associated cost, or duty, generally in the form of taxes. This is borne by the members of that society. It is clear that the elevation of any desire to the status of a 'right' will likewise carry a cost, and this may or may not be something which the members of society are willing to assume. Any rights-based claim needs to be carefully weighed against its cost, whether this is social or economic.

These issues came to the fore in 2002, when one of the HFEA's rulings was challenged by a patient and her clinician. It was suggested that the patient's reproductive rights were being infringed by restrictive regulations.

Case 26: Mrs H[12]

Mrs H married when she was 41. She and her husband wished to start a family but found that they were unable to conceive a child. In May 1996, Mrs H consulted Mr Taranissi, the Medical Director of the Assisted Reproduction and Gynaecology Centre.

Between June 1996 and July 2000, Mrs H underwent eight IVF treatment cycles, in each of which three of her embryos were replaced in the uterus. This was in keeping with a ruling by the HFEA which stated that 'no more than three eggs or embryos should be placed in the woman in any one cycle, regardless of the procedure used'.

Mrs H failed to become pregnant, and Mr Taranissi subsequently wrote to the HFEA requesting that the ban on implanting more than three embryos should be waived in this case. He claimed that for an older patient such as Mrs H, the risk of a multiple birth was non-existent but that the use of more than three embryos might give her a reasonable chance of conceiving. Mr Taranissi planned to maximize Mrs H's chances of becoming pregnant by implanting five embryos.

The court ruled that the HFEA's restriction was justified, and that the case did not justify a relaxation of the regulations.

Mr Taranissi argued that it was appropriate to make an exception to the rule in the case of this particular patient, based on her specific treatment needs. At the time the case went to court, Mrs H was 47 years old. Clearly, there was only limited reproductive time left for her. She had suffered a number of gruelling and tragic setbacks in her attempts to have a child. A miscarriage in 1984 had been followed by a diagnosis of blocked fallopian tubes. Surgery had been attempted but was unsuccessful. The eight failed cycles of IVF which had followed, all involving the implantation of three embryos, had demonstrated that her chances of success with any further cycles looked bleak. Because of this, Mrs H and Mr Taranissi were convinced that to achieve any possibility of success, it was vital to increase the number of embryos which could be implanted.

Mr Taranissi was clear that he did not wish to challenge the general validity of the HFEA's three-embryo rule. However, he believed that in exceptional circumstances, it should be possible for a clinician, in consultation with his patient, to override this provision. His letter to the HFEA stated:

> It is my strong belief that the replacement of more than three embryos in this particular case is not just medically essential and professionally sound, but is the minimum required in the discharge of my professional responsibility towards this patient.

Before going on to consider the human rights implications of this case it is important to consider the reasoning behind the HFEA's insistence that no more than three embryos should be implanted at any one time. Natural conceptions result in twins in around one out of every eighty pregnancies, while the chances of twins being born from an IVF pregnancy is around one in four. Largely, this is due to the implantation of more than one embryo. This is usually undertaken not with the specific aim of producing multiple births, but with the intention of increasing the chances of at least one of the embryos successfully implanting and surviving to birth.

For infertile patients, the prospect of having several children at once might seem a doubly or trebly joyful event. However the birth of twins and triplets is far more risky than the birth of a single child.

Complications of multiple births[13]

For the mother:

- Increase in anaemia because the foetuses make excessive demands on the woman's ability to provide nutrients in the womb
- Pregnancy-induced hypertension and pre-eclampsia (potentially fatal)
- Gestational diabetes
- Greater risk of surgical intervention or forceps assisted delivery
- Increased maternal mortality rate
- Mothers of multiples are more likely to experience social isolation, depressive illness, and child abuse

For the child:

- Low birthweight
- Increased risk of infant mortality
- Congenital malformations
- Premature birth
- More likely to require artificial ventilation
- Risk of intracranial bleeding
- Cerebral palsy

Social and economic costs:[14]

- Costs to the NHS per singleton: £3313
- Costs to the NHS per twin: £9122
- Costs to the NHS per triplet: £32,354
- Multiple pregnancies after IVF are associated with 56% of the direct cost of IVF pregnancies, although they represent less than one-third of the total annual number of maternities in the UK

Babies with low birth weights are far more likely to suffer from serious lifelong health problems, as are premature babies. The incidence of neonatal death in these infants is also a serious issue. The risks associated with multiple births increase with the number of babies, so having triplets is more risky than having twins, and so on. If Mr Taranissi had implanted five embryos in his client there was a possibility that she could have found herself pregnant with quintuplets.

Multiple pregnancies where the mother is carrying too many children for her own or their safety can sometimes be 'reduced'. This is a euphemism meaning that some of the developing foetuses are aborted in order to give the remaining foetus or foetuses an improved chance of surviving the pregnancy. This is an extremely undesirable way of dealing with a multiple pregnancy, both in terms of the emotional and physical effects on the mother, and because it imposes a risk on the remaining foetus or foetuses.

These risks of multiple births, coupled with the undesirability of foetal reduction, formed the basis of HFEA's reasoning in establishing a fixed limit of three embryos for transfer. One might well ask—since the risks associated with multiple births are so great—why three embryos should be allowed, when allowing only one embryo to be implanted should effectively solve the problem? The answer to this was that the three embryo limit had been carefully deliberated by the HFEA as a compromise between permitting the optimum chance of success in an IVF cycle for infertile patients, while keeping the risks of multiple births within reasonable parameters. With improvements in implantation techniques, and greater awareness of the risks of multiple births, the limit has since been reviewed and is now set at two. It is likely that with further improvements it will eventually be reduced to one.

However, if it is the case that a patient has a *right* to a particular treatment, it may be that these risks are not sufficient reason to withhold that treatment. When Mrs H's case came to judicial review, the HFEA was obliged to consider the possibility that certain articles of the European Convention on Human Rights and Fundamental Freedoms might undermine its position.

Article 8

1. Everyone has the right to respect for his private and family life, his home and his correspondence.

2. There shall be no interference by a public authority with the exercise of this right except such as is in accordance with the law and is necessary in a democratic society in the interests of national security, public safety or the economic well-being of the country, for the prevention of disorder or crime, for the protection of health or morals, or for the protection of the rights and freedoms of others.

Article 12

Men and women of marriageable age have the right to marry and to found a family, according to the national laws governing the exercise of this right.

The scope of these rights is not absolute: the right to privacy is circumscribed by whatever laws exist in the area, and considerations of security, as well as conservation of resources. It is also circumscribed by considerations of harm to others. The right to marry and found a family, on the other hand, is restricted only according to national laws.

For the HFEA, then, it was necessary to demonstrate that its intervention in the matter of the number of embryos to transfer was warranted by one of the considerations outlined above. In the first instance, the question was whether the HFEA's stance was in accordance with domestic law. Plainly, this was indeed the case, since the law in the UK dictates that the HFEA must determine what constitutes suitable practice and issue guidelines accordingly.

The question of whether the interference pursued a legitimate aim also had to be addressed. The aim in this case was to avoid the health risks associated with a multiple pregnancy and multiple birth, which might affect both the mother and any children born as a result of treatment.

It was also necessary to consider the cost implications of these health risks. As shown above, the burden of relieving the health problems which ensue from fertility treatment places a financial strain on the NHS and thus on society as a whole. It should be noted that around 75% of the (extremely expensive) IVF work carried out in Britain is paid for privately by the patients who undergo treatment.[15] On the other hand, the provision of care for the pregnancies which result from treatment usually devolves upon the NHS.

Given the huge expense of providing treatment for desperately ill premature or underweight babies, as well as the possible lifelong health implications such children face if they survive, these burdens on the NHS have to be considered when weighing up individuals' supposed right to have a child at any cost against the question of who will actually meet that cost.

Thus the protection of health and the safeguarding of public health resources underpinned the HFEA's stance on this matter—both of which considerations are specifically allowed for in the rights legislation. Finally, in considering proportionality, the HFEA balanced the importance of the need for interference against any detriment to the rights of the individual patient, ie Mrs H. The HFEA carefully considered such statistics as there were relating to pregnancies and multiple births among women of that age, both in this country and abroad. It found that the birth rate was very low indeed. Given the previous history of Mrs H's fertility treatment, it seemed extremely unlikely that any future treatment would succeed however many embryos were transferred. Because of this, the plan to increase the number of transferred embryos was regarded as being likely to have little significant effect in increasing the patient's chances of achieving a pregnancy.

If, against the odds, Mrs H did become pregnant, the risk of a multiple pregnancy would be proportionally higher if five embryos had been transferred rather than three. In cases of older women who become pregnant as a result of fertility treatment, the risk of a multiple birth still applies and this is all the more threatening at an older age. In the context of the extremely low likelihood of pregnancy for this patient the possibility of a marginal improvement in the possibility of achieving a pregnancy was outweighed by the attendant risk of a multiple pregnancy if five embryos were transferred.

The results of the judicial review upheld the HFEA's decision, and Mr Justice Wall[16] found that the HFEA's reasoning could not be described as irrational. He observed that this was an area of rapidly developing science in which judicial review had a limited role to play. Disagreements between doctors and scientific bodies in this pioneering field were inevitable. Mr Taranissi was therefore obliged to uphold the rule of the HFEA Code of Practice, as the patient's human rights could not justify a relaxation of the rules in these circumstances.

This outcome shows that while human rights legislation has an influence on the HFEA's reasoning, the purported existence of a human right does not necessarily override the interest that society has in regulating fertility treatment. The HFE Act in the UK is an expression of the democratic process, which has concluded that fertility treatments require a system of licensing and regulation. The HFEA is the appointed body to carry out these tasks.

The HFEA's decisions may be legitimately challenged if it is seen to act or make decisions which are beyond the sphere of the powers given to it by Parliament. However, it is also sometimes challenged when acting within its appropriate remit, as in this case. Inevitably, not all clinicians or patients will agree with the HFEA's position with regard to regulation and they may sometimes claim that their rights are being contravened. However, again, the ruling here shows that such rights are not regarded as being absolute.

A perverse incentive?

As suggested, the cost of fertility treatments can be extremely high. Fertility specialists are reported to be among the highest-earning medical professionals,[17] earning more than even the most successful plastic surgeons. The immediate costs of fertility treatment are usually borne by the patients themselves (NHS provision is patchy). This being the case, there may be another incentive for patients to feel that multiple births are not an unwelcome consequence of IVF techniques. If a patient has two—or even three—children as a result of one IVF cycle, it may provide a 'complete family' in one go.

Achieving this would spare the patient the prospect of further IVF cycles, with all the attendant costs, physical and emotional burdens, and the uncertainty of success. In effect, the result of a multiple birth could be regarded as a financial saving—getting three for the price of one. Because of this apparent bonus of multiple births for fertility patients, it is all the more important that prospective parents are aware of the risks involved.

The fact of the financial incentive is unfortunate;[18] however, the fact remains that patients may assume these risks despite being made aware of the dangers. People embarking on fertility treatments have to make a significant commitment. Once this commitment has been made, it may be tempting to focus on maximizing the chances of a pregnancy whatever the dangers. This fact in itself is an important consideration in maintaining and enforcing regulatory restrictions.

The downside of rights and European law

Rights legislation is often assumed to be a good thing. It protects the interests of individuals against the interests of the State. It is surely desirable that individuals should be protected against unwarranted intrusions into their liberty or privacy. Nevertheless, interpreting rights legislation is extremely difficult. It is not always obvious what constitutes an unwarranted infringement: local or national laws may supersede supposedly universal 'rights' but the circumstances in which this is justified may be open to debate.

Court cases are costly and time-consuming. Rights rhetoric tends to focus on the individual, perhaps to the detriment of legitimate social concerns. For all these reasons and more, the impact of human rights legislation on the regulation of ART—as well as other areas—may bring some negative consequences.

The application of European law in domestic courts has caused (and is likely to continue to cause) a lowering of national standards to meet the lowest to be found in Europe. For example, where a procedure or treatment is illegal in the UK, but is legal or unregulated in one or more other European countries, an individual can simply avoid the UK's restrictions by travelling abroad. This has given rise to what is termed 'procreative tourism'.

To some degree this might be interpreted as a welcome liberalizing phenomenon which allows greater numbers of individuals to satisfy their desires. However, viewed from another angle it can be seen to set in motion a worrying trend towards the acceptance of risk and a rejection of precautionary principles. As demonstrated in the discussion of the case of Mrs H above, much of the regulatory framework which surrounds fertility treatment in the UK arises from social concerns about the medical and economic consequences of unrestricted practices in this field. Where the HFEA's regulations can be overcome simply by taking a short aeroplane journey to a different country, this undermines the purpose and integrity of domestic legislation.

The startling effect of the combination of human rights legislation and European Treaty principles, in the UK at least, is to dissolve restraints. The relevant articles of the EC Treaty are Article 59 (now 49): '... restrictions on freedom to provide services within the Community shall be progressively abolished . . . in respect of nationals of Member States who are established in a State of the Community other than that of the person for whom the services are intended' and Article 60 (now 50): 'Services shall be considered to be "services" within the meaning of the Treaty where they are normally provided for remuneration . . . "services" shall in particular include . . . activities of the professions'.

Silently and unnoticed, the application of human rights legislation and EC Treaty principles invites people to countenance risks deemed unacceptable in their home countries.

In the UK, the development of legislation concerning issues that pose risks to public health or the environment is very well advanced and sophisticated. Every regulatory decision is taken in the knowledge that it may be challenged in court. Human rights legislation and the EC Treaty provisions tend to undermine and devalue this legacy. The application of the Human Rights Act 1998 has fostered an overweening concern with individual rights and liberties in a legislative environment that was previously carefully balanced so as to attain a fair weighting between public health requirements and the desires of the individual. The EC Treaty principles of freedom of movement of goods and services, and the right to seek medical services abroad, may in the last analysis undo all the careful regulatory constraints applied in the UK.

It is questionable how far a regulatory body can usefully operate if its jurisdiction may simply be overridden or circumvented by travelling abroad. It also seems intrinsically unjust that, in effect, the HFEA's regulations may thus be seen to apply only to those who are not affluent enough to travel abroad to seek treatment. This leads to a situation in which the fertility treatment of the poor is subject to regulatory restrictions, while the rich are unhindered in their reproductive endeavours, however risky or reprehensible these may appear to be.

Rights and the precautionary principle

To an extent these problems mirror a tension between the interests of individuals to seek their own ends, and the application of the precautionary principle. In essence, the precautionary principle entails that when doubt exists as to the safety of a practice, authorities should err on the side of caution. It is assumed that the status quo is likely to be less harmful or risky than the implementation of an uncertain or untested technology. When the HFEA was deliberating on the risks of using frozen and thawed eggs for use in fertility treatments, its decision to prohibit the use until their safety had been demonstrated was an exercise of the precautionary principle.

In February 2000 the European Commission published a Communication stating that the precautionary principle should be brought into play only when a specific potential risk has been identified and evaluated. Then, it states, decision-makers have to respond according to the degree of risk that they perceive to be acceptable to the society they govern.

These risks have to be weighed against the freedom and rights of individuals, industries and organizations. The response in the case of risk should be proportionate, consistent, and open to revision. One of the interesting results of this is that on the basis of this Communication, so-called 'unknown unknowns' cannot be factored into the analysis. Where a new technology is under consideration, according to the statement of the European Commission, it is only acceptable to act on the precautionary principle in relation to risks that have been *identified* as being connected with the new technology.

Clearly when new technologies or procedures are being evaluated, particularly in the field of fertility treatment, there may be limited scope to form an accurate picture of all the possible outcomes. However, the procedure might still be regarded as involving too great a degree of uncertainty to be permissible—at least according to the precautionary principle as adopted by the HFEA.

Technology throws many new challenges our way: reproductive cloning, artificial gametes, and genetic engineering may all be on the horizon. In the case of these ground-breaking technologies, it may be impossible to pinpoint all the associated risks. For example, wholly unpredictable consequences could potentially result from attempts at cloning. Such a view was certainly shared by the National Bioethics Advisory Commission, which cited 'unknown risk' in its assessment of safety issues as part of the justification for recommending a ban on human cloning.[19]

If the precautionary principle can only be adopted where risks are identifiable, the onus falls on those who are concerned about the dangers to point out where these dangers may lie. In cases where this is not possible, untested procedures may be forced into use. Whether this is desirable or not is open to debate. From a libertarian view, upholding individuals' rights to assume risks, and rejecting restrictions, the European Commission's Communication would be welcomed. However many believe that technological developments should be applied cautiously, and on the assumption that they may bring in their wake unforeseen challenges and dangers.

Homogeneity and national integrity

The provision of services in relation to reproductive technologies and embryo research presents difficult moral considerations with respect to human rights and in the context of the European Union. Until recently, it was accepted that Member States could legitimately differ in terms of what was regarded as being morally acceptable. The integrity of different countries' legislative processes were seen as matters to be decided according to the needs and concerns of their own populations. However, the emergence of guidance such as the European Commission's statement of the precautionary principle necessarily thwart the ability of Member States to maintain this integrity.

Guidance may be obtained from the well-established and highly developed set of individual human rights prevailing throughout European jurisprudence. But it will be necessary to go further and deeper to develop principles resting on the understanding of the goals and methods of medicine and biological research. We do not lead our lives in isolation, but as members of communities, large and small. To these communities we owe certain duties as members and citizens. Some of these duties reflect the fact that others may rely on us in certain circumstances, not only to avoid causing them harm but also to take positive steps to assist them. Drawing the boundaries of this web of claims and responsibilities is a complex task.

Decisions as to where to draw these boundaries will reflect local, national, and cultural concerns that may not necessarily be shared by other countries. In trying to reach a consensus on human rights and apply it as broadly as possible, it may be found that the Articles designed to protect individuals become vague and insubstantial. Conversely, the homogenizing trend may erode previously erected safeguards, leaving people more, rather than less, vulnerable.

CHAPTER 7
Deconstructing The Family

Modern scientific advances in the field of ART have a unique capacity to challenge our assumptions about what constitutes a 'normal' family. Can the notion of 'family' survive in an environment where surrogacy is available and parenthood can be split into separate genetic, social and gestational components? Where embryos can be created without the need for sex, and can be frozen, to be thawed perhaps years after the death of the individuals from whose sperm and eggs they were created?

It has been suggested that reproductive technologies in general may put unusual strain on family, and especially parental relationships:

> ... the [...] shift from an unconditional, given relationship between parents and children to a conditional and chosen relationship [is] considered a problem, since chosen relationships might not have the same force with respect to imperfect (specific) obligations parents have towards their children: to love them, to give them some continuity and stability.[1]

In fact, these techniques which are challenging our concepts of family life to such a degree have been creeping up on us gradually for some time. The first schism occurred when the contraceptive pill became widely available in the 1960s. This made it possible effectively to separate sex from childbearing. This was followed by a growing acceptance of cohabitation and illegitimacy. Sex and childbearing had been split apart, and now the necessity of marriage for family life had come into question. With the continuing development of ART, together with the availability of donated gametes, same sex couples can seek to reproduce, and at some point in the future, perhaps, artificial gametes and/or reproductive cloning may allow these couples to circumvent the need for gamete donors. Artificial gametes might also provide a means for an individual to reproduce without the need for a partner.

In view of these developments, perhaps a concept of 'family' is no longer needed in reproductive technology decisions. The erosion of social unease regarding unmarried sexual partners and illegitimate offspring may suggest that current assumptions and moral convictions about what constitutes a family will also be overturned. Are there any values which were inherent in the previously held convictions about the necessity of family structures, or were these assumptions based on mere prejudice and lack of imagination? Is modern disregard of 'the family' anything more than a gesture towards what seems to be fashionable?

These questions are not merely academic or whimsical musings; they form the basis of some of the most controversial pieces of regulation in the field of reproductive technologies. Fertility clinics are obliged by law to consider the welfare of any children who might be born as a result of their interventions, and this legal obligation specifically refers to the family setup.

The Warnock Report stated:

> . . . the interests of the child dictate that it should be born into a home where there is a loving, stable, heterosexual relationship and that, therefore the *deliberate* creation of a child for a woman who is not a partner in such a relationship is morally wrong . . .

and

> . . . we believe that as a general rule it is better for children to be born into a two-parent family, with both father and mother . . .[2]

Section 13(5) of the HFE Act stipulates that

> a woman shall not be provided with treatment services unless account has been taken of the welfare of any child who may be born as a result of the treatment (including the need of that child for a father) and of any other child who may be affected by the birth.

The final wording of this clause was the subject of some debate. Parliament voted down a number of more stringent requirements, including a suggestion that only married parties should be eligible for fertility treatment. Nevertheless, Mary Warnock and her committee clearly felt that the traditional family is indeed preferable to other setups. More recently, speaking as UK Prime Minister, Tony Blair affirmed his belief that marriage is a good thing, and single parents who have chosen to have children without forming stable relationships are wrong.[3]

The question might be asked why marriage itself should make a difference? Marital breakdown is common, and divorce has become socially acceptable, so the mere existence of a marriage cannot by itself guarantee two parents for a child over the longer term. (A belief that parents should be married might be taken to imply that same sex couples should not be encouraged to become parents. But same sex couples can enter into civil partnerships which confer legal rights and responsibilities almost identical to those of marriage.)

But despite the high divorce figures, the odds of a cohabiting couple with a young child splitting up are more than twice that of a married couple of equivalent age, income, educational attainment, and ethnic group.[4] Of course, it might be argued that these cohabiting couples would be just as likely to split up if they got married. Perhaps, whether consciously or unconsciously, they have avoided marriage because they are uncertain of their commitment to each other.

Alternatively, the explanation of these figures may lie in the fact that, apart from whatever emotional and psychological bonds marriage may foster, the financial, legal and practical entanglements involved make it more difficult for partners to extricate themselves from unhappy marriages than from unhappy cohabitations. Taking a more positive view, one could speculate that marriages are more stable than cohabitations because the ceremony really does confer an added sense of commitment and cohesion which tends to support a happy relationship.

Whether or not this should speak in favour of marriage is debatable. At any rate, it provides an additional explanation as to why marriages may be more stable than ad hoc partnerships. As Tony Blair's statement above shows, although political and moral attitudes toward marriage and family life may have changed in many respects, the idea that children should be raised by two—preferably married—heterosexual adults still seems to hold sway. But even within this traditional framework, controversies can arise.

Donated gametes

Donated gametes, especially donated sperm, have been in use for many years. It is one of the lowest-tech forms of fertility treatment, and one might assume that it would therefore be less controversial. However, this is far from true. There are vocal witnesses who testify that it is extremely harmful to be deprived of a link with one's genetic parent.[5] In the UK until recently, those who donated sperm or eggs to infertile couples could do so in the knowledge that they would remain anonymous. Naturally, this meant there was no opportunity for offspring to form relationships with their donor parents, which was a source of severe distress to many donor-conceived children.

From 1 April 2005, the UK law on gamete donation was changed. Donors no longer have the right to remain anonymous: offspring can seek their genetic parent after the age of 18.

Case 27: Locating gamete donor parents

New developments in technology and DNA testing have allowed for some interesting approaches to questions of paternity. In 2005, a 15-year-old boy who had been conceived with the sperm of an anonymous donor succeeded in locating his genetic father.[6]

The boy found an online DNA testing service designed to help people piece together their family tree. He sent them a swab test for analysis. The donor was not on the company's database, but the sample was a close match with two people who were. It suggested that all three had a relative in common.

Using this information, the boy conducted further online research and eventually managed to identify the man with whose sperm he had been conceived.

Whatever regulatory processes are in place, enterprising individuals may be able to circumvent them. Men who donated gametes prior to the change in the law may have read about this case with great interest. Even if men could still donate anonymously, ongoing anonymity cannot be guaranteed. The fact that the child in this case was still legally a child is also striking. The law only allows offspring to seek their donor parent after the age of 18, but clearly there are some children who are not prepared to wait this long.

The problem of gamete donation is a strangely asymmetrical issue in some respects. Growing up without knowing one's genetic father is not a new development. Some studies have suggested that many 'fathers' who are bringing up children together with their wives or partners, are not in fact biologically related to their supposed offspring. In some cases, of course, the man is aware of this fact, but in a surprisingly large number of cases, he remains oblivious of the situation unless some combination of circumstances brings it to the fore. (There is debate as to what the actual prevalence of non-paternity is, and it has been suggested that the figures have been overestimated.)[7]

This highlights the oft-iterated fact that while a mother knows for certain whether a child is hers or not, the father generally has to take the mother's word for it. Clearly, then, if there is a problem involved with not knowing one's genetic parent, it may affect many more people than just those who were conceived using the sperm of a donor.

Different views on anonymity

Some of the issues related to donated gametes were discussed in the House of Commons Science and Technology Committee's Report *Human Reproductive Technologies and the Law,* which was published in March 2005. The evidence gathered for this report involved a discussion from a number of perspectives, including representatives from fertility clinics and adults who had themselves been conceived from donated gametes. The difference in weight attached to the anonymity of gamete donors by the different parties was extremely telling.

One of the specific fears associated with the removal of donor anonymity was that there would be a reduction in the number of gamete donors. It was speculated that potential donors would be put off by the idea that future offspring might one day contact them, throwing the donor's life into disarray. Indeed, the number of gamete donors in the UK is very low, and this fact has been highlighted by a number of clinics.

However, fertility clinics have a vested interest in maintaining the numbers of donors, since many of their treatments rely on the availability of donated gametes. Given this, it has been suggested that perhaps clinics' perceptions are skewed in their own favour.[8] They would prefer that donors have the right to remain anonymous, because it is assumed that this will encourage more donors to come forward.

Those who are born as a result of anonymous gamete donation may not share these views. Some believe that children have a 'fundamental right to know the identity of their biological parents'. Others feel that 'every individual has a right not to be deliberately deceived or deprived of information about essential aspects of their personal history by the public authorities'.[9]

A disparity between the views of the clinicians and those of some of the individuals affected by donor conception is perhaps unsurprising given the different motivations of those involved.

It is important to note that it may not necessarily be the loss of a relationship with the genetic parent alone that causes distress to donor-conceived children. Often the trauma may be exacerbated by having been deceived. Donor-conceived children are likely to assume that the people who bring them up are their genetic parents, unless informed otherwise. The failure of parents to correct this assumption can be looked on as being deliberately deceptive, especially if the truth emerges suddenly due to unforeseen circumstances. There is some evidence to suggest that if parents are open about their use of donor gametes from the start of the child's life, the families—and particularly the children themselves—find it much easier to deal with[10] than those who suddenly discover the truth or are told, whether deliberately or accidentally.[11]

A further point to be made here relates to the importance placed on biological relationships. The term 'biological parents' in the statement above is used to differentiate between 'social parents' who, although they bring the child up, are not genetically related to that child, and the genetic parent who donated the gametes.

Gamete donation and adoption may have the effect of divorcing the social from the biological where parenthood is concerned, and this is often felt to be problematic. However, many reproductive possibilities, including artificial gametes, surrogacy, and cloning, further subdivide the concept of parenthood, allowing for distinctions *within* the definition of 'biological' parent.

Babies with two mothers?[12]

Most of the genetic information in a cell is held in its nucleus. However, cells also contain what is known as 'mitochondrial DNA' outside the nucleus. The function of mitochondrial DNA is not fully understood. It comes only from the mother, and is passed on to offspring through the maternal egg cell. Some women are known to have genetic abnormalities in their mitochondrial DNA. If these women reproduce they may transmit these abnormalities to their offspring, causing disease, disability, and sometimes death.

It has been suggested that this can be overcome by removing the nucleus from the affected mother's fertilized egg and inserting it into an empty egg cell donated by a woman whose mitochondrial DNA is normal. The resulting child would be free of disease. A child born of such a procedure would carry genetic material from two women as well as the man whose sperm fertilized the egg ... but would either, or both, of these women be properly regarded as the biological mother?[13]

If there is a right to know one's genetic parents, would this mean that a child conceived in this manner had a right to know *all* of those concerned in its genetic makeup?

This question reveals the complexity behind the apparently simple concept that everyone has exactly two biological parents. New technologies may well affect how relationships with genetic parents are perceived in the future. Perhaps the concept of genetic parenthood will come under such pressure from these developments that it will ultimately give way to accommodate the broader possibilities of those who can contribute to the genetic makeup of a child. Alternatively, those who support and nurture the child may be seen as his or her 'real' parents. At any rate, it does not seem altogether clear that the notion of biological or genetic parenthood is entirely straightforward, and this complexity is likely to increase over time with further technological developments.

Surrogacy

Case 28: Helen Beasley[14]

When Helen Beasley, a 26-year-old British woman, entered into a surrogacy arrangement with a Californian couple, she did not imagine that her desire to help another couple to have a baby would end in recriminations and court proceedings.

What had gone wrong? Instead of the single foetus that the commissioning couple, Charles Wheeler and Martha Berman, had requested, it transpired that Ms Beasley was carrying twins. Wheeler and Berman argued that under the terms of their agreement, Ms Beasley would be obliged to abort one of the foetuses if she should become pregnant with twins.

However, Ms Beasley felt unable to go through with the abortion which Wheeler and Berman had scheduled for her thirteenth week of pregnancy, and she argued that in fact there had been a verbal agreement that any such termination would have to take place before the twelfth week. Additionally, she was concerned that this kind of selective termination can harm the foetus which is to be retained, and increase the likelihood of miscarriage, making it more probable that both foetuses would be lost.

Wheeler and Berman withdrew from the surrogacy agreement, leaving Ms Beasley carrying now not just one but two unwanted foetuses, neither of which she was prepared to abort. Ms Beasley decided to try to have the twins adopted, but found that she could not do so since under Californian law it is the commissioning parents who are regarded as having legal rights over the offspring.

The fact that some women are prepared to carry and give birth to children, only to hand them over in return for money is thought by some people to be outrageous. The practice seems to rank with prostitution. It degrades an act which should surely be a reflection of love into a mere commercial transaction. Of course, there is a distinction to be made here between a surrogacy arrangement which involves payment and one which does not. However, even where payment is not at issue, the idea that a woman will hand over 'her' child to someone else seems to grate against our preconceptions of motherhood, and of the bond that gestation entails.

When the payment is involved, it is all too easy to draw analogies not just with prostitution, but with the buying and selling of human beings themselves. Understandably, this is an uncomfortable thought. We regard the days of slavery as being far behind us. Yet if babies are transferred from one adult to another in return for money, are we embarking on something just as reprehensible? Is this in any way more acceptable than the direct sale of embryos or gametes?

Apart from these problems, there are serious legal difficulties which arise from the technological possibility of separating gestational, genetic, and social motherhood in this way, as can be seen from the case above. Global attitudes and legal responses to surrogacy are extremely varied. In the US, surrogacy contracts are regarded as legally binding. Provided the contract has been properly drawn up, the commissioning parents (those who pay the money) are regarded as the default legal parents of the children. This means that surrogate mothers in the US who change their minds are not in a position to claim any right to keep the child; the courts are unlikely to favour them.

In the UK, on the other hand, it is the woman who gives birth to a child who is regarded as its legal mother, whether or not she is also the genetic mother. This means that if individuals in the UK enter into surrogacy arrangements with each other, these contracts will not be regarded as binding. Even if agreements have been meticulously drawn up and money has changed hands, the 'ownership' of the child(ren) by the commissioning parents will not be recognized.

The reason why the actual performance of surrogacy—that is to say, carrying a child for another person—is not illegal per se in the UK is that it was felt such a law would be impossible to enforce. There is a reluctance to allow the law to probe too deeply into our social and sexual activities. Enforcing a law against surrogacy would require very stringent probing indeed.

The technological requirements for donor sperm insemination are fairly basic, and do not necessarily entail clinical expertise, or treatment in a hospital. Insemination of the prospective father's sperm can be performed at home with some basic kitchen equipment. A surrogate who uses her own eggs may thus perform the operation herself with relative ease. Alternatively, if donated sperm is used it could be obtained from one of the various unregulated donor sperm providers which exist in the UK and elsewhere (although if the sperm was stored, it would fall under the jurisdiction of the HFEA). Clearly, if the surrogate were not using her own eggs, more complicated techniques would be necessary and clinical involvement would be unavoidable.

Since many unforeseen events can occur, as exemplified in the case described above, surrogacy can be a risky business. Multiple births, the birth of damaged or diseased babies, changes of heart in the surrogate mother, may all present a legal problem which does not admit of any easy solution.

But in some cases surrogacy arrangements work well. Unexpected complications do not always arise, and where each party receives the side of the bargain they were expecting, should there be anything further to say about it? Perhaps there should. Even though legal problems, or disputes over the 'ownership' of the children may not always arise, there is the bald fact that surrogate mothers give birth to children they do not want to bring up themselves. Could this cause psychological problems, or feelings of abandonment in the children?

There is an obvious analogy here with adoption. It is not uncommon for adopted children to feel resentful at having been forsaken by the women who gave birth to them. A strong urge to find out the story behind the adoption is also frequent. These feelings can be pervasive even though the adoptive family atmosphere is loving and nurturing.

Carrying a child and giving birth to it are surely among the most profound experiences that a human being can have. At no other time are we so intimately connected with another person. The level of commitment involved in carrying and giving birth to a child is extreme. Mothers must go through nine months of pregnancy, often experiencing discomfort which may range from morning sickness to life-threatening conditions such as pre-eclampsia. At the end of the pregnancy, they must endure the process of labour and childbirth, or undergo a caesarean operation, both of which options involve considerable risks.[15,16]

Perhaps because of this level of commitment, it is assumed that women who undergo pregnancy and childbirth have a high stake in the outcome, in other words that they must *really* want their child. Because of the pain involved in childbirth we tend to think of mothers who have given birth as being connected with their children in a very primeval kind of way. Who would go through this experience unless they longed for a child? Having gone through this experience, what kind of woman would be willing to give up her child?

Yet giving birth does not always guarantee a flow of motherly love towards the newborn infant. Undoubtedly, giving birth is a hugely significant experience, but it would be hasty to assume that it is either necessary or sufficient for parent/child bonding. No doubt many a mother-to-be is unwilling, ignorant, or pressurized, and did not freely choose to subject herself to this emotional bond.

Again, it is interesting to compare this with adoption. The birth mother of an adopted child also undergoes the experience of childbirth and all its pains, yet we do not necessarily regard parents who give up their children for adoption as being culpable. Adoption is often described in terms of wanting to give the child the best chance at life, and this is regarded as rather to be commended than condemned. Of course, in adoption the transaction is not expressed in financial terms, although in reality financial considerations are frequently a primary factor in motivating mothers to give their children up for adoption. Granted, the mothers do not receive payment as such, but their future lives are relieved of the financial burden that the child would have represented. Is this really so different from paid surrogacy?

There is a further significant difference which needs to be addressed: we usually consider mothers who relinquish children for adoption as doing so reluctantly, having accidentally become pregnant, or having experienced an unexpected change in circumstances. We assume that birth mothers do not deliberately decide to become parents in order to relinquish their children, and they part with their children in tearful anguish, or so our imaginations portray it. This moral difference sets adoption apart from the deliberate choice undertaken in surrogacy. However, it is interesting to question the soundness of this reasoning. After all, if mothers giving their children up for adoption suffer anguish, why should this make it more morally acceptable than surrogacy? Arguably, it should make it less so.

Perhaps what makes us feel uneasy about the concept of surrogacy is not so much the idea that money is a motivating factor, but rather that the commissioning party is willing to persuade a woman to undergo the physical and emotional effort of bringing a child into the world, perhaps to suffer unanticipated pain at the point of relinquishing that child. There are some well-publicized accounts of surrogates who have undergone a change of heart and suffered terrible anguish at giving up 'their' children—just as much, in fact, as women who give children up for adoption.

It may be impossible to know how a woman will feel after going through the experience of childbirth. Someone who embarks on a surrogacy contract with a woman risks the possibility that her attitude toward her pregnancy may change, and that she may suffer terrible emotional turmoil at giving up the child to whom she has given birth. Of course, some or even most[17] surrogates perhaps do not experience this sorrow, but the possibility of this emotional response is perhaps what makes us feel that there is something disconcerting about the idea of forming a legal contract to give birth to, and then relinquish, a child.

The impact on the children themselves should also be considered. Are the children of surrogates psychologically or emotionally damaged by the fact that the women who gave birth to them were willing to part with them, perhaps in return for money? As with other psychological and emotional issues related to the impact of reproductive technologies on children, some of the proof of these claims is still awaited. It is not always easy to follow up on children born to surrogates, partly because of the shadowy role of surrogacy in the UK legal system. However, there have been studies which suggest that young children born to surrogate mothers do not suffer any obvious psychological damage.[18, 19] An earlier study indicates that there are no additional physical risks to children born to surrogates.[20]

Adverse psychological or emotional effects may manifest themselves later on in life, and there is little data to assist us here. However, perhaps the best approach to this problem is to continue to be aware of the need for follow-up where possible. Before clear evidence has emerged, it is impossible to condemn surrogacy specifically on the grounds of trauma to the child: as yet there are simply no data available either to support or to discredit such a claim.

In the absence of such evidence, perhaps the precautionary principle should be adopted. This would require that we should not embark on new procedures since we cannot be sure they will not be harmful, and the only way of obtaining empirical evidence is to allow the procedure to go ahead on an experimental basis. It is hard to experiment in reproductive medicine; either one has a baby or one does not. Perhaps taking a gamble on procedures which may fundamentally affect a child's health or well-being is not acceptable. There is some strength in this kind of reasoning, especially in the context of reproductive technologies which do not, after all, save endangered lives.

But the emotional drive of prospective patients is a powerful force in propelling forward the application of new technologies, and the corresponding evolution of family structures. The extent to which we should control or constrain this force is a matter of continuous debate. Generally, the precautionary approach is applied with relation to physiological risks to children born as a result of reproductive technologies. It is less frequently applied to the psychological and emotional risks, or to broader risks to society in general. This may be because such risks are far harder to identify and quantify.

Perhaps the best we can do in these circumstances is to monitor the children, and make sure that any adverse effects are noted, and used in determining whether the State should take any more active role in the future in discouraging surrogacy. However, it should be recognized that allowing practices to continue on this tentative and experimental basis creates a precedent which could make it extremely difficult to legislate subsequently, even on the basis of adverse results.

Single parents: no need for a father ... or a mother?

With the development of artificial insemination and other fertility treatments, it has become possible for women to become mothers almost independently of men. This development has been welcomed by some as an expansion of women's reproductive freedom, but it is viewed with anxiety by others. The HFE Act does not forbid single women from accessing fertility treatments. However, it specifically requires clinicians to consider the need of the child for a father. This has given rise to criticism:

> The requirement to consider whether a child born as a result of assisted reproduction needs a father is too open to interpretation and unjustifiably offensive to many. It is wrong for legislation to imply that unjustified discrimination against 'unconventional families' is acceptable.[21]

Responses from Government ministers suggested that this clause might indeed be removed from the HFE Act[22,23] (although there are certainly some who will oppose this). But would the removal of this clause send a strong message to men that they are not needed? Perhaps it is no wonder that there are so many men who walk out on their parenting and financial responsibilities if the paternal contribution to a child's upbringing is regarded as being so unnecessary. In some ways it could be said that use of IVF in this way treats men as nothing other than genetic donors, not as real people at all.

In the context of family law, ensuring children's continued contact with their father after divorce, or relationship break up, is regarded as being extremely important. This may be explained partly as a way of ensuring a degree of stability and continuity for children when families are fragmented after divorce. However, it also represents a more fundamental conviction of the importance of knowing our origins, and maintaining relationships with those who have begotten us. A change in the law to remove the statutory requirement of consideration of the need for a father would be inconsistent with many of these principles.

The idea that children have a psychological need for an identifiable father is also reinforced by the requirement (as of April 2005) that sperm—and egg—donors should no longer be anonymous. Realistically, this has a far greater impact on male gamete donors, since egg donation is relatively rare. Thus, again, the importance of fatherhood seems to be very much built into the law.

Historically, there were many additional reasons for the law to recognise the importance of fathers. Fatherless families were morally, socially, and economically problematic. Wives who lost husbands and children who lost fathers were the objects of pity and charity and often faced destitution if such charity were not forthcoming. Unmarried women who had children were subject to the utmost moral censure and the taint of illegitimacy blighted the lives of the children themselves.

The stigma of being an unmarried mother or an illegitimate child has all but vanished in most respects. Many couples choose not to get married, and their offspring are no longer regarded as pariahs. But has the specifically economic role of the father changed? Women are now able to earn broadly comparable salaries with men.[24] This means that, should one parent take sole responsibility for the family's economic welfare, it could feasibly be the woman. Of course, in practice this is not usually the case and in our society it is not uncommon for both parents to work. A single income in today's economy is often insufficient to provide for family life. This means that the absence of a supplementary income may indeed adversely affect single parent families, but—in theory at least—this could apply to the lack of either partner.

In reality though, the vast majority of single parents are women. And while single modern women can, in principle, earn comparable wages to men, the necessity of caring for their children often severely circumscribes their earning potential. The result is that even if single parenthood is no longer stigmatized for explicitly moral reasons, single mothers and their children still represent some of the most disadvantaged members of our society.

Yet this is not always the case. Most single parents are women and most single parents are also financially disadvantaged, but let us consider an example which breaks the mould on both counts.

Case 29: Ian Mucklejohn

Ian Mucklejohn is a man who is used to success; used to getting what he wants in life.[25] A wealthy businessman, he had reached the age of 58 when he realized that he had failed to achieve one of the most important things in life: having a family. Not one to be daunted,

Mucklejohn set about rectifying this situation. Perhaps other men—and women—faced with the difficulty of producing a family in double quick time might have hastily formed a relationship, had children, and hoped for the best.

Mr Mucklejohn tackled the problem from a different angle. He was sceptical about the chances of finding the right woman to be the mother of his children at this relatively late stage in his life. Rather than embark on an ill-judged relationship, he decided to use his financial power to fulfil his desire for offspring.

He located two women in America to help him in his quest. One of them provided her eggs to be fertilized with Mr Mucklejohn's sperm. The other woman agreed to be implanted with the fertilized eggs, carry the foetuses and give birth to the babies. The costs involved were in the order of £50,000. Mr Mucklejohn is now the father of triplet boys.

For those of us who have been lucky enough to have had a loving mother, the association between childhood and a mother's love may be inextricably linked in our minds. Could it be that Mr Mucklejohn's sons will suffer a fundamental lack in their lives, due to the absence of maternal input and affection? Whatever the answer to this question, it seems odd that while the need for a father is mentioned in law, the need for a mother is not.

Does this mean that while single women's desires are constrained, single men are entitled to pursue their parental ends in whatever ways suit them? It seems implausible that those involved in debating and drafting the Act believed children need fathers more than they need mothers. But if this is so, the omission of the need for a mother seems to rest on an assumption that this simply goes without saying: of course a child needs a mother. The need for a father, on the other hand, needs to be spelt out.

There is an interesting historical parallel to be made here: during the years in which male homosexuality was illegal, female homosexuality was not. This was because Queen Victoria thought it self-evident that women would not indulge in such horrors. Likewise those who drew up the HFE Act presumably never envisaged the possibility that single men might seek fertility treatment, so the question of a child's 'need for a mother' was never raised. Yet single men can and do seek to reproduce, as the case of Mr Mucklejohn demonstrates.

Mr Mucklejohn deliberately chose to procreate in the knowledge that he would be a lone parent. As he had no partner, these children effectively have no social mother; nor do they have regular contact with their genetic or gestational mothers. Of course, the presence of two parents is no guarantee of good outcomes for children, and the offspring of such parents are not immune to all or any of the problems which can afflict children in general. However, statistical evidence that children do better on balance when born to couples is an important consideration, even if marriage or cohabitation in itself cannot guarantee this.

From a practical point of view, it seems fairly evident that having more than one pair of hands is desirable when bringing up children.[26] One wonders how Mr Mucklejohn fared with triplets! Related to the practical issue is the financial question mentioned above. Single parents may find themselves struggling to care for the family as well as maintaining their careers. Achieving both may be difficult or impossible, resulting in the family being forced to rely on meagre State benefits.

Where there are two parents, there is more flexibility for the division of responsibilities, and at least the possibility of two incomes. If there is only one parent, things are necessarily different. But are they worse? It is often claimed that the children of single parents are more likely to struggle at school and suffer from a number of other social problems, and there is empirical evidence to support this fact.[27,28]

However, when compared with studies of single parents who have consciously chosen their status as single parents, an interesting discrepancy emerges. Children in these circumstances do not seem obviously disadvantaged, and do not seem to struggle in the same way as children in 'conventional' single parent families.[29,30] (It should be mentioned here that these studies refer to the children of single women. There is little data on families headed by a single father.)

Where single parenthood is planned, the parent is likely to have weighed up the financial issues before proceeding: the pregnancy will not come as a shock, and will have been incorporated into the budget. By definition, planned single-parenthood does not come about by accident. And crucially, the traumas which are likely to be implicated in unplanned single parenthood (such as relationship breakdown or the death of a partner, all of which may affect the children and remaining partner, whether directly or indirectly) simply do not apply where single parenthood has been deliberately chosen.[31]

Nevertheless, there is some research which suggests that the input of a father specifically is beneficial or even essential to a child's optimal development.

That is to say, if the research is correct, a child may suffer from the lack of a father per se: statistically his or her prospects might not be as good as those of a child brought up in a household where the father is present.[32] (Again, the assumption here is that the mother is present by default.)

The financial concern is also important. If lone parenthood is statistically linked to economic hardship, we might question whether it is right to bring into existence children whose lives may be blighted by all the problems which financial insecurity can bring.

In theory this could be addressed by increasing the level of State financial support available to single parents so that they and their families are not economically disadvantaged relative to other groups. Alternatively, greater practical contributions could be made—such as the provision of childcare—to enable single parents to re-enter employment. (Both of these options would, of course, involve a cost to society.)

However it is also necessary to consider the fact that some lone parents, such as the wealthy Mr Mucklejohn, are quite clearly not victims of financial hardship. While, statistically speaking, many single parents struggle financially, lone parenthood in itself is not necessarily a guarantee of this. The financial ability of prospective parents to support a child is something that can be addressed separately from their single status. This is specifically recognized in the HFEA Code of Practice.

Independently of these concerns, the idea of deliberately bringing a child into the world with only one parent—of whatever sex—is viewed by some as being obviously socially and morally wrong. Two parents, one male and one female, have for thousands of years been the necessary means of bringing children into the world. A mother and father jointly bringing up children may provide an example of complementary roles, the interaction between the two sexes, a role model of each sex, and two sets of relatives. There is exposure to adult conversation, and to an adult relationship that does not centre solely round the child. Two sets of interests and life experiences enrich the family sphere, and there is support available should one parent become incapacitated or be absent.

In sum, there is give and take to be witnessed, a micro-society of which the child is just one member. And part of the joy of having a child in the 'natural' way is perhaps the feeling that the child is the outcome of its parents' love for each other. A child is not produced simply to satisfy an individual's desire to be a parent. Children embody their parents' relationship in a unique way.

If we hold this view, then seeking to have a child to satisfy our own interests, without reflecting the shared experience of another person, could be seen as selfish in the most literal way. However, the ideal of having children as an embodiment of a loving relationship does not always hold true. Perhaps relatively few parents—even in happy relationships—specifically reproduce with the idea of 'incarnating' their love. The truth may be far more prosaic.

Parents have children for all sorts of reasons, or for no reason at all. Even in this age of reliable contraception, surprising numbers of pregnant women describe their pregnancies as being unplanned.[33] And of course, even where two parents are present, children may still suffer. Indeed, they may suffer specifically because the parents *are* both present. Smouldering resentment, extra-marital affairs, fights, and constant sniping may result when parents stay together for the sake of the children. In such circumstances it is questionable whether two parents are necessarily better than one.

There are a variety of conflicting views on these issues and inevitably the law cannot accommodate them all. The welfare of the child clause is often spoken of as though it militated against the creation of 'unconventional' families. But this is not the case. Mr Mucklejohn's treatment was undertaken in the US, which is relatively unregulated, but even in the strictly-regulated UK, there is nothing in the HFE Act which would necessarily have prevented Mr Mucklejohn or any other adult from receiving treatment solely on the basis of his or her single status.

Clinicians have some latitude in their interpretation of the welfare of the child clause, which is fleshed out in the HFEA Code of Practice:

3.2.2 The centre should carry out a risk assessment in relation to each patient and their partner (if applicable) before any treatment is provided. The assessment should be carried out in a non-discriminatory way. In particular, patients should not be discriminated against on grounds of gender, race, disability, sexual orientation, religious belief or age.

3.3.1 Those seeking treatment are entitled to a fair assessment. The centre is expected to conduct the assessment with skill and care, and have regard to the wishes of all those involved.

3.3.2 In order to take into account the welfare of the child, the centre should consider factors which are likely to cause serious physical, psychological or medical harm, either to the child to be born or to any existing child of the family. These factors include:

(a) any aspect of the patient's (or, where applicable, their partner's) past or current circumstances which means that either the child to be

born or any existing child of the family is likely to experience serious physical or psychological harm or neglect. Such aspects might include:
 (i) previous convictions relating to harming children, or
 (ii) child protection measures taken regarding existing children, or
 (iii) serious violence or discord within the family environment;

(b) any aspect of the patient's (or, where applicable, their partner's) past or current circumstances which is likely to lead to an inability to care for the child to be born throughout its childhood or which are already seriously impairing the care of any existing child of the family. Such aspects might include:
 (i) mental or physical conditions, or
 (ii) drug or alcohol abuse;

(c) any aspect of the patient's (or, where applicable, their partner's) medical history which means that the child to be born is likely to suffer from a serious medical condition;

(d) any other aspects of the patient's (or, where applicable, their partner's) circumstances which treatment centres consider to be likely to cause serious harm to the child to be born or any existing child of the family.

3.3.3 Where the child will have no legal father, the centre should assess the prospective mother's ability to meet the child/children's needs and the ability of other persons within the family or social circle willing to share responsibility for those needs.

These welfare requirements are designed to minimize the risk of harm, but they do not rule out any particular type of family structure. The case-by-case approach attempts to avoid the injustice of systematically barring certain groups from accessing treatment. In contrast to the approach taken by many other European countries, British fertility law fixes no upper age limit and sets no legal restrictions based on marital status or sexuality.

A parliamentary report published in 2005 complained bitterly about the welfare of the child clause on the grounds that it is discriminatory:

The welfare of the child provision discriminates against the infertile and some sections of society . . . it should be abolished in its current form.[34]

However, a more recent parliamentary report indicated a growing consensus that this clause, or some form of it, is a vital part of the legislation:

We support the approach taken in the 1990 Act towards the welfare of the child and the positive shift in the HFEA guidance towards a risk-based approach.[35]

Non-discrimination

If we take the need to avoid discrimination seriously, perhaps there should be no restrictions on who can access fertility treatments. Since the fertile can reproduce without having to meet any State-imposed criteria, any legislative interest in the reproductive choices of those who seek treatment might seem discriminatory. But is this an acceptable line to take? We should also remember here that fertile people do not have to pay to reproduce. Therefore if we are truly non-discriminatory, fertility treatments should be provided free of charge to those who require them.

Sheila McLean has argued that:

> whilst the state may have no general duty to facilitate reproduction through technology or the supply of a partner, once facilities are provided—for example through *in vitro* fertilisation and surrogacy programmes—to deny access on the grounds of sexuality is to infringe the right on a discriminatory basis.[36]

There is obviously a degree of truth in this: once access to the technology is allowed to some, restrictions on its use will necessarily be discriminatory. (Although, as noted above, it is important to bear in mind that access is *not* denied on the grounds of sexuality; discrimination on these grounds is specifically ruled out in the HFEA Code of Practice.)

The fact is that all provision of medical treatment is discriminatory simply because we do not have sufficient resources to provide everything everyone wants or needs.[37] Sheila McLean's statement suggests that the important issue is whether we discriminate on clinical or social/behavioural grounds.

The dispute over the welfare of the child clause turns not solely on its claim to protect children from harm. Those who take issue with the clause believe that it is founded on what they perceive as moral convictions rather than facts. Supposing this were true, it might indeed seem disturbing. Limiting access to scarce resources might justifiably be based on proof of harm to children, but can it be justifiably based on moral beliefs? We do not usually sanction the parcelling out of medical resources based on the basis of this kind of judgement. Criminals, for example, are supposed to receive no less preferential healthcare treatment than other members of society. Discrimination as to who should receive treatment is—in the UK—supposed to be based squarely on judgements of medical need.

In fact, however, broader value judgements do seem to be making their way into clinical decisions on resource allocation. As financial pressures increase, the primary care trusts responsible for allocating treatments are seeking new ways of conserving funds. In recent times, we are frequently enjoined to take responsibility for the condition of our health. To access certain treatments, prospective patients may need to show that they are controlling their weight, or that they are not smokers.[38]

The moral component of such rationing is highlighted when one considers that behaviour which is not regarded as morally culpable, but is equally dangerous or foolhardy, is not taken as a reason to deny treatment. Thus, for example, an obsessive jogger who puts excessive strain on his joints and requires costly knee surgery might receive treatment where an overweight patient would be turned away. The boundary between clinical and moral judgements turns out to be surprisingly permeable.

But despite this, perhaps it is still possible to try to ensure that discrimination as to who should receive fertility treatment is based purely on clinical grounds.

If treatment were offered only to those who are medically infertile, the problem of making value judgements about the preferred family type might evaporate. This kind of approach would conserve resources (by restricting the group of eligible claimants) and maintain a strictly clinical approach free from the taint of moral opprobrium. It might also incidentally preserve traditional family structures.

But when one starts to examine the 'true' nature of infertility it appears to be a condition which—perhaps more than many others—is socially constructed. A 'normal' heterosexual couple experiencing fertility problems may receive treatment even though only one member of the couple is technically infertile. A woman in perfect reproductive health may be eligible for treatment with IVF because her partner has fertility problems. A woman in this situation could be regarded as being 'contextually' infertile even though she has no underlying clinical disease. In the same way, a single man or woman, or people in same sex partnerships, might also be regarded as being contextually infertile.

Moreover, unexplained infertility is extremely common. One of the 'diagnoses' of infertility is simply a failure to conceive after a certain period of having unprotected sex. Often, there may be no identifiable medical cause at all, but this does not mean that patients are denied treatment. The links between clinical infertility and access to treatment are weakened further in the case of fertile heterosexual couples who seek fertility treatments and PGD to avoid passing on diseases to their offspring. Again, neither patient is necessarily infertile, yet treatment is not withheld from them. In addition, where fertility treatment is used in order to create a 'saviour sibling' for an ailing child, there is no connection with clinical infertility.

Infertility is often a 'social' disease and access to treatment cannot always be decided on the basis of purely clinical concerns. Reproduction is an area of intense social and political interest. The Government's concern over future children's welfare extends into all of our lives. Supplements are added to the food we eat in order to protect the health of the children we may have. Women are bombarded with information and instructions on how to ensure they do not put their future children at risk, for example with warnings against alcohol consumption, stress, smoking, and so on.

Applying selection criteria to fertility treatment is not necessarily out of keeping with this attitude of concern for the next generation. Indeed, it might seem perverse to ignore the possibility of preventing harm to the children born through the use of reproductive technology. Fertile women are the focus of more or less direct efforts to influence or control their reproductive choices, so why should this interest not be extended to those who require fertility treatment?

It seems fair to say that infertile people and single and same sex parents should be subject to the same concerns and restrictions that affect everyone's reproductive decisions. This might mean that if there is thought to be a risk of serious harm to their child, they may be refused treatment. But the question of how to establish the risk of harm may be difficult to answer.

The role of empirical data

The distinction between preventing harm to children and making implicit or explicit value judgements about family type is not always as clear as one might think. Those who support 'traditional' families often bolster their arguments with suggestions that such families are in fact better for children pragmatically, socially, and economically. Conversely, those who adopt new interpretations of the family are equally likely to cite evidence showing that children are not harmed by 'unusual' family structures. Empirical data relating to the wellbeing of children in unconventional families is thus a hotly-contested prize, and both sides claim it as their own.

Same sex parents

Case 30: Barrie Drewitt and Tony Barlow

In 1999, long-term couple Barrie Drewitt and Tony Barlow decided to pursue their long-held desire to start a family.[39] After they were refused the opportunity to adopt a child in the UK, they located an American surrogate to carry their future offspring. They obtained a donated egg from another woman, and this was fertilized with the sperm of one of the men. In due course, Barrie and Tony became the proud fathers of twins.

The prospect of a child being brought up by 'two mothers' or 'two fathers' may seem disturbing to some, partly because of the putative effects on the child, but also because of the challenge to the accepted family form. The law does not restrict access to fertility treatments on the basis of sexuality, but some clinics are less inclined than others to treat same sex couples and single people. (This was particularly true in the decades immediately following the development of IVF.)

The use of empirical research in the context of same sex couples and their access to fertility treatments highlights the complex relationship between subjective values and supposedly objective facts. It has been suggested that children brought up by homosexual partners will suffer gender confusion, and that their own sexuality will be influenced.[40] (One might ask, in a society that supposedly repudiates homophobia, whether it should be a problem even if it were the case that children brought up by same sex couples were more likely to be gay.)

However, other research suggests that these fears are unfounded.[41] A number of studies have suggested that children brought up by same sex couples do not suffer the problems that were predicted.[42,43] On the other hand, it has been argued that the research on which such claims are based was deficient in, for example, having inadequate comparison groups, non-random samples, too small samples, or unreliable measurements.[44]

Where conflicting empirical evidence emerges, it is not clear how to proceed. This problem is not unique to the question of same sex parents, of course, but applies to any area where there is dispute including many of the issues discussed above.

The questions involved here often have a far deeper value to many of those who feel strongly about these questions. Some of those who oppose the creation of 'alternative' families might do so even if empirical data proved irrefutably that these families are safe for children. Likewise, those who advocate a liberal non-discriminatory approach might accept some family structures even if children were shown to be harmed by them. Avoiding discrimination would be seen in these cases as having moral primacy. The conflict then is not always simply to be resolved by pointing to empirical evidence; in many cases this is a red herring.

The effects that 'new' family types may have on children are often unknown. Where there is uncertainty, again a dichotomy of reasoning emerges. Those who favour the traditional family may adopt the precautionary principle, arguing that a child who has never existed cannot be harmed, so when in doubt treatment can be withheld. Those who prefer the libertarian approach insist that proof of harm occurring to children is required if access to treatment is to be denied. Yet in some circumstances, it is simply not possible to provide proof.

Where proof is lacking, underlying values and beliefs come into conflict. The growing weight attached to non-discrimination is an instance of one moral conviction superseding another. The way we view this rise and fall of beliefs may vary. Some people conceive morality as a continuum of improvement and enlightenment. Thus, changes in values and assumptions are seen as desirable and progressive. Others may see different moral views as representing a constant flux of social and cultural pressures, not progressing anywhere in particular but merely reflecting the changing cultural and environmental contexts from which they emerge.

Either way, when views differ, it is hard to establish which, if any, should be enforced by legislation. And empirical data cannot always provide an answer.

Changing values

With the passage of time, the assumptions which underpinned the Warnock Committee's overt belief in the desirability of two heterosexual parents have been challenged. To an extent, developments in reproductive technology itself have contributed to the evolution—or erosion, some might call it—of value judgements connected with family life. During the time that has passed since the publication of the Warnock Report, huge changes have taken place:

> ... established boundaries were decomposing. The nature and stability of the family and marriage were waining [*sic*]; the openness and acceptability of heterosexual partnerships without marriage and homosexual relationships without moment were waxing; and established social and legally enforced gender roles and stereotypes, both outside and inside the home, were being recast in the long shadows of feminism ... The institutional security ... of established ethical concerns was under fundamental assault from feminisms, the collapse of traditional theological canons and the emergence of new questions in the methodology and epistemology of ethics.[45]

If anything, this shows that debate is likely to be constant. Any statute is merely a milestone on the way, as it cannot solve ethical problems even though it may solve legal ones. On balance perhaps the HFE Act has got it about right. The HFEA Code of Practice which expands on the welfare of the child clause, has been adjusted year by year in line with changing social concerns. Inevitably, further changes will be made in the future.

Our laws cannot reflect the sheer variety of moral customs and beliefs held by the population. In general, then, the law enforces what might be termed the lowest common denominator of morality: the Millian principle, which states that freedom should be curtailed only where there is risk of harm to others. This is what the current regulatory framework seeks to do, while recognizing the challenges inherent in identifying and defining what constitutes risk.

Clearly, this is not identical with everyone's perception of morality, but—in theory at least—differing value systems can operate within this legal context, without necessarily being enforced by it. Accordingly, the law around fertility treatments avoids proscriptions based on moral assumptions about the appropriateness of certain family types. Rather, the most basic kind of risk-avoidance strategy is adopted.

Of course, if we have strong values and beliefs, we may wish that we could enforce them, or at least persuade others of their validity. In modern society perhaps the only way we can be sure of finding fruitful ground to impart our own values is by inculcating them in our families, in particular in our children. No wonder we place such importance on family life. But what actually constitutes family life is unclear. Is parenthood conferred by genetic relationships, or by the care and energy expended by adults on their children?

Whatever the answers to these questions may be, it is evident that the family as a concept, perhaps as an entity, is undergoing radical changes, some of which are reflected in new legislation about civil partnership and cohabitation. However family structures might be interpreted now and in the future, what is important is that they will still provide what is necessary to nurture our children and allow them to flourish.

CHAPTER 8
Embryonic Stem Cells and Therapeutic Cloning

The scientific, social, and ethical fallout from the successful development of *in vitro* fertilization techniques extended far beyond questions related exclusively to reproduction. People became aware of the existence of a new class of entity: embryos created and stored in the laboratory. What was the moral status of such embryos? For scientists, the research possibilities were immense. When the idea of performing research on these embryos began to filter into the public consciousness, a split in the nation's attitude manifested itself. To an extent, this split mirrored existing concerns about abortion. However, the scope for embryo experimentation added an entirely new dimension to these arguments. It seemed impossible that both sides of this divide could be appeased by any regulatory framework, yet this was the task which lay ahead.

Britain prides itself on being a country with a robust regulatory framework that allows research on embryos, while ensuring that such research is carefully monitored and overseen, and that the status of the embryos is protected. However, it was not always inevitable that this would be the case. Around the time of the publication of the Warnock Report, there was a vociferous and well-organized body of people who were willing to put time, effort, and rhetoric into fighting for legislation against all research on human embryos. In February 1985, the Unborn Children (Protection) Bill received its second reading and was debated in parliament. As Michael Mulkay has said, this:

> proved to be a resounding victory for the opponents of embryo research and confirmed the worst fears of the scientific community.[1]

Although the scientific community was not surprised at the success of the bill as such, it was both shocked and dismayed at the sheer scale of the opposition to embryo research (the vote was 238 to 66). The opponents of embryo research had marshalled their forces effectively and successfully and had simply outdone their pro-science rivals.

Following on from this, there was a reaction from scientists and journals such as *Nature*, making an effort to address the fears which had led to the Bill's triumph, and attempting to engage with the debate rather than simply assuming that their research was justified. Thus, it began to seem that scientists themselves had a role to play in convincing a sceptical public of the need for embryo research if they wished to carry out such research without being subject to criminal proceedings. They could no longer remain passive, assuming that their work would be unquestioningly condoned by the public; they began to realize that, as Mulkay puts it:

> all research has to be justified to the satisfaction of the lay community and its parliamentary representatives.[2]

At the same time, those politicians who supported embryo research became aware that they would need to organize themselves as efficiently as their opponents had done; they would need to employ equal rhetorical ingenuity and—crucially—make a convincing case for allowing research, based on the benefits which the proposed research could bring. Since the opponents of embryo research seemed to have arrogated to themselves the emotional spectrum of argument, the pro-science lobby made use of the weapons which remained, and which seemed so appropriate for their cause: rationality, adherence to facts, statistics, and so on, coupled with a positive and aggressive approach.[3]

Thus a polarity arose in the debates on this issue, which persists to this day. The Department of Health's 2006 *Review of the Human Fertilisation and Embryology Act* states that its aim in reviewing the legislation (including among other things legalized research on embryos) is 'to pursue the common good through a system broadly acceptable to society'.[4] But where views are so polarized, is it really possible that any legal framework that allows embryo research can be said to be 'broadly acceptable to society'?

The polarization of the debate around embryo research has also been mirrored in other areas. Public feeling against genetically modified (GM) food crops revealed a degree of mistrust not only of scientists, but crucially also of politicians, and of the institutions which were set up to evaluate the safety of GM foods. This public mistrust has been compounded in the aftermath of the BSE crisis and the foot and mouth epidemic, and with the fear of avian flu. Misgivings of this sort are often portrayed by scientists and politicians as stemming from an emotional, irrational, risk-averse, and under-educated proportion of the public.[5] For example, the then Prime Minister Tony Blair, in his speech to the Royal Society in 2002, described perceptions of the GM debate as having been 'completely overrun by protestors and pressure groups who used emotion to drive out reason'.[6]

On one side of the divide, we are presented with the rational, progressive, pro-science lobby, while on the other side, those pitted against them are portrayed as ill-informed lay people, and those with religious convictions or conservative leanings. This side of the debate has lost out overall; its views are regarded as being obsolete, and the political power which backed them up has been vanquished. Clearly, then, it is not just science itself that has advanced swiftly since the pre-Warnock days. There seems to have been a sea change, perhaps matching the change of government, and a new mood of aggressive political optimism with regard to scientific advances.

In the new political era, science makes for credibility. Evidence, statistics, figures, quantified risks, and so on are all acceptable considerations for science, and to an extent even non-scientific disciplines have been forced to adopt the language of science to produce quantifiable evidence, if they wish to be taken seriously. Pro-science policy-makers are described as seeking 'the maximum freedom for individual adults in the absence of demonstrable harm'.[7] Accordingly, the intuitive or emotional component of discourse in many spheres has been repudiated or discredited in favour of more scientific, dispassionate, or evidence-based language.[8]

In some measure, this dispassionate approach might be seen to be justified. Many people who were aghast at the idea of test-tube babies may have been reacting to a 'yuck factor'. Emotional and intuitive responses made us hesitate at the idea of breaking with the 'natural' requirements for reproduction. But the passage of time has shown us that Louise Brown and the thousands of IVF babies who have been born since then are not monstrous; we have not plummeted down a slippery slope into Frankenstein-type machinations or the creation of a subhuman underclass. Perhaps, then, this demonstrates that emotional or intuitive moral concerns are simply fears of the unknown, that they are indeed irrational and should not be allowed to come between scientists and the objects of their research.

There is a flux in the acceptability of certain arguments and the desirability of an empirical approach. Neither the conservative, emotive approach nor the libertarian progressive approach necessarily have the upper hand for very long. Over the past few decades there has been an increasing tendency for formerly widely-held beliefs, such as the importance of marriage, the unacceptability of homosexuality, and the illegitimacy of abortion, to be viewed as old-fashioned, paternalistic, and unreasonably restrictive. Simultaneously, a kind of repugnance for the basis on which these morals were held has manifested itself in the discourse around these issues. Political correctness and the avoidance of discrimination are the concerns of the day, giving rise to a libertarian approach that requires the maximum freedom for individuals, and puts the onus on those with misgivings to provide conclusive evidence of why contested procedures are wrong.

The fact that the general acceptability of emotional beliefs has diminished so rapidly may well in turn diminish the credibility of this kind of belief in general (that is to say, beliefs which do not purport to be founded on the basis of empirical evidence). This may be all the more true given that the hard core of people who do hold such beliefs are often seen as strident, irrational, and unamenable to the persuasions of rational argument. For a society to become aware of its moral evolution may be unnerving, and these facts may well serve to make individuals and policy-makers alike more wary of trusting 'moral intuitions', 'gut reactions', or judgements based on emotion, especially in a legislative context.

Yet for anyone who wants to push back the barriers (whether in the interests of science or commerce) a return to the Warnock Report sounds a cautionary note. The Report repeatedly cited the need for boundaries, for controls and restraints on science, and the relationship between emotion and ethics. It is against this backdrop that the ongoing arguments around embryonic stem cell research are being played out.

What are embryonic stem cells?

Before looking more deeply at the arguments related to embryo research, let us briefly address the question of why this kind of research is of such interest to scientists.

When an egg cell is fertilized, it begins to divide. These first few divisions create cells identical to one another up till around the third or fourth day after conception. At this stage the fertilized egg (or blastocyst) may contain over 100 identical cells. These cells are what is known as 'totipotent'. That is to say, if any one cell was isolated from the others, it would in theory have the capacity to form a whole new embryo as well as the placenta and umbilical cord. However, the period of totipotentiality is limited to the first three or so days after conception, while all the cells are still identical. From this point onwards, the cells start to differentiate. Some are destined to make up the umbilical cord and placenta, while others—in the inner cell mass of the blastocyst—will make up the body of the foetus.

The cells from the inner mass of the blastocyst are pluripotent: they have the capability to turn into any cell in the human body (but not, it is thought, the placenta and umbilical cord). They are also 'immortal', in that they have the ability to renew themselves indefinitely in laboratory conditions. However, in the normal course of the development of a human being, the cells become more and more specialized until nearly all of them have assumed fixed characteristics as, for example, heart, liver, lung, or skin cells. Once this has happened, it is thought that they can no longer be 'reprogrammed', and are no longer 'immortal' in the way that pluripotent stem cells are.

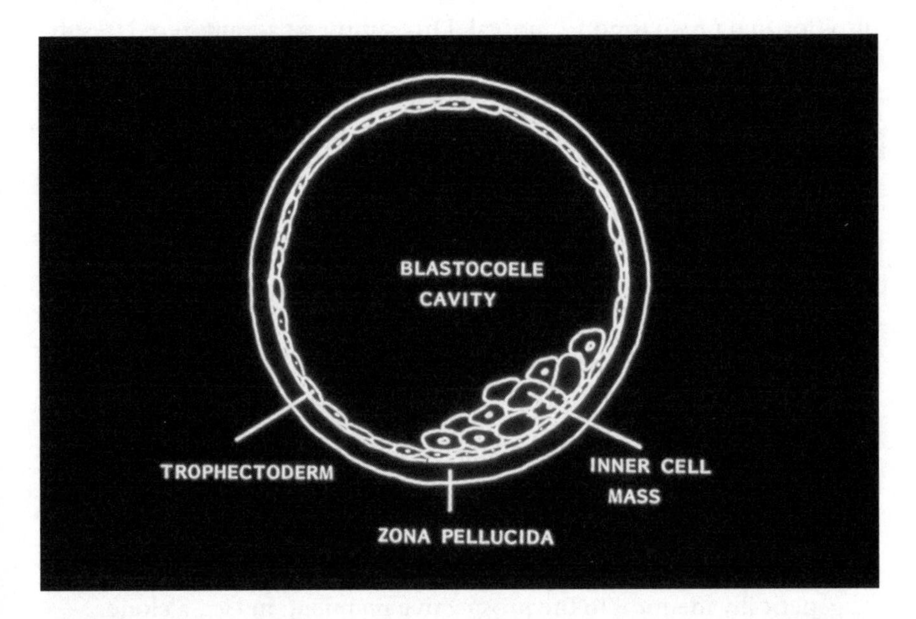

The blastocyst is a hollow ball. The trophectoderm will become the placenta. The inner cell mass gives rise to the foetus itself, and it is these cells from which embryonic stem cells can be obtained.

Embryonic stem cell lines are derived from the pluripotent cells in the inner cell mass of the blastocyst before they have had time to undergo this differentiation. Because they are removed at a time when their stem cell properties are intact, they retain the potential to become any cell of the body—*in the laboratory*. It is this capacity which makes them so exciting to scientists. Suppose that we could grow heart, brain and liver cells in laboratories. Could we use these cells to cure the diseases that afflict these organs? Could we grow whole organs themselves?

Currently, organs for transplantation are in desperately short supply. Many people wait in vain for organs to become available, but even when a suitable organ is located, receiving a transplant is not always the end of the problem. Our immune systems recognize our own DNA and tissue types, and any foreign material which enters the body is vigorously attacked and destroyed. The reasons for this are fairly evident. If foreign matter enters our bodies, it brings with it the risk of infection and cross contamination. We need the vigilance of our immune systems to protect us against invading parasites, microbes, germs, bacteria, fungi, and viruses, all of which can pose a serious threat to our health.

However, this vigilance works to our detriment when we are reliant on a donated organ for our survival. Drugs must be administered to subdue the immune system after an organ has been transplanted, as without these drugs, the new organ would be recognized as 'foreign'. However, the immuno-suppressant drugs themselves (which must be taken for the rest of the recipient's life) have some serious side effects, and can lead to an increased likelihood of cancer.

Because of this, even where organs are available for transplant, they do not present a perfect cure. However, in conjunction with therapeutic cloning, it might be possible to develop tissue which is genetically identical to that of the recipient. How would this work?

1. A donated egg would have its nucleus removed.
2. The nucleus from an adult's body cell (such as a skin cell, for example) would be injected into the egg.
3. The egg would be stimulated by electricity into starting to develop into an embryo. If allowed to develop fully, this embryo would be genetically identical to the prospective recipient: in fact, a clone.
4. The development of the embryo would be arrested, and the inner cell mass removed.

Possible Research Applications of Cloning Which do not Create Identical Human Beings

- Advances in medical research are very difficult to predict but animal and human cloning could improve basic understanding of biology.
    - Ageing process—DNA damage in adult cells
    - Cancer—could lead to therapies

Replacing Cells, Tissues and Organs

- Cloned cells would reduce risks of rejection

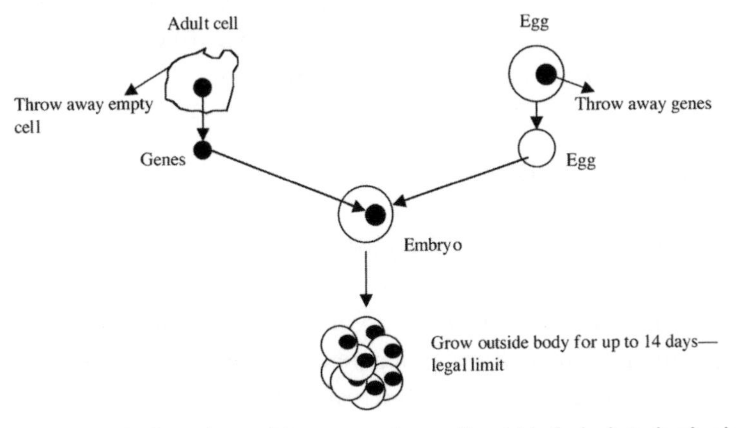

It might in the future be possible to treat embryo cells outside the body to develop into:

- cells such as bone marrow
- tissues such as skin—more complex
- organs such as kidneys

Other Research

Currently, the deprogramming of adult cells requires the incubating of the nucleus in an unfertilised egg—the creation of a cloned embryo. Further research may lead to new ways of converting adult cells into other cell types.

What differences would there be if cells, tissues and organs could be grown without the need to first create an embryo as a source of cells?

Therapeutic cloning

5. The cells collected would be cultured into stem cell lines.
6. The stem cells would be differentiated into cells of the type required, eg heart cells. Each of these cells would have the same genetic make-up as the prospective recipient.
7. The tissue would be transplanted into the recipient. The risk of rejection, and consequent need for immuno-suppressant drugs would be obviated because the tissue would be genetically identical to that of the patient.

This process is highly complex, and it is likely to be many years before it becomes feasible, if indeed this ever happens. The processes by which embryonic stem cells begin to differentiate are still poorly understood. Any therapeutic potential in embryonic stem cells (ESCs) relies on our being able to understand and control these processes.

Foetal stem cells

Stem cells can also be obtained from foetuses. However, unless obtained very early in foetal development, such cells are unlikely to retain the same capacity to differentiate as embryonic stem cells and thus they are not necessarily preferable from a therapeutic point of view. Foetuses are not terminated specifically for such research, but their use may still raise ethical concerns.

Case 31: Mr Tricarico

In 2005 the BBC reported the story[9] of a man who was dying of Duchenne's Muscular Dystrophy—a progressive disease which destroys the muscles of sufferers. Most of those affected with the condition will die in their twenties. In 2005, Stefano Tricarico was 20 years old, and he was desperately seeking a cure for the disease.

Mr Tricarico had identified a clinic in the Ukraine, which purported to offer stem cell therapies which could offer the chance of a cure. Treatment would cost large sums of money, but to Mr Tricarico it was literally a question of life and death. He embarked on a campaign to raise funds to enable him to undergo the treatment.

Other scientists were sceptical about the Ukrainian clinic's claims, suggesting that the scientific evidence was simply not there. The procedure also raised ethical concerns. Was the clinic really offering a realistic chance of a cure, or simply taking advantage of a desperate patient?

The fact that the 'cure' relied on the use of cells obtained from aborted foetuses was also regarded as a potential problem. Since the treatment was experimental, there was also the risk that something could go wrong. It is not always clear how stem cells will behave once transplanted into patients' bodies.

However, Mr Tricarico assured critics that his condition had improved as a result of the treatment. For him, he believed that it had been worth it.

This case shows that in unregulated environments, striking claims are being made as to the achievements of scientists working with stem cells. However, most experts remain unconvinced of the therapeutic benefits of the treatment offered to Mr Tricarico, and the burgeoning market in foetuses and foetal cells has also been viewed with alarm.

A less contentious use of foetal stem cells involves the removal of cord blood from the umbilicus of newborn babies. This blood is rich in stem cells, and its use is well documented. Since the blood is taken only from the cut umbilical cord, it poses no danger either to the newborn baby or the mother.

Adult stem cells

While research into embryonic stem cells has been continuing apace, parallel efforts have been focussed on adult stem cells. It well known that adults' bodies also contain stem cells, and that these can be used to generate further cells. However, there are a number of reasons why adult stem cells are regarded as having less therapeutic potential than ESCs. Most obviously, adults' bodies consist largely of cells which are specialized to perform one particular function for a certain period and then die. Identifying the relatively few stem cells among these normal body cells and removing them would be likely to be painstaking and invasive.

Furthermore, the stem cells obtained from adults are thought to be more highly differentiated than those from embryos. They may only be capable of producing one or two kinds of related cell, whereas ESCs have much wider potential (although this has been disputed by some scientists). Adult stem cells also seem to lack the 'immortal' properties of embryonic stem cells. Despite this, it seems evident that there is much more to learn about adult stem cells. Over the past few years a number of publications have reported the discovery of adult stem cells in areas where they were previously thought not to exist. The possibility of using adult stem cells may be appealing largely for the reason that it would circumvent many of the ethical issues which so encumber ESC research, and which have entirely prevented this kind of research in some countries. However, for the moment it seems likely that research along both avenues will continue.

Ethical concerns

The legal status of the embryo and of embryo research were settled in the HFE Act. The Act was based on the findings of the Warnock Report which concluded, after weighing the arguments on both sides, that the embryo 'deserved respect'. How this is interpreted remains an open question, and the accordance of respect to embryos failed to satisfy those who were vociferously opposed to any embryo research at all.

After it emerged that the cloning of mammals had become a possibility, the HFE Act was extended to permit embryo research into serious disease, as well as into infertility. Parliamentary approval was secured, in part, by the assurance that the ethical issues would be examined by a Select Committee. The House of Lords Select Committee on Stem Cell Research was chaired by the Bishop of Oxford and reported in 2002. The Report concluded that there was a clear scientific case for continued research in embryonic stem cells. It was hoped that the technique of cloning would allow embryonic stem cells to be developed which would be genetically identical to the donor. This might eventually enable the creation of individualized replacement cells for people suffering from certain diseases. The House of Lords Committee also suggested that research on embryos should continue and was not unethical, but that embryos should not be created specifically for research purposes unless there was a demonstrable and exceptional need that could not be met by the use of surplus embryos generated in IVF treatment.

The use of cell nuclear replacement was approved with the same limitation. The Committee stated that it trusted the HFEA with its legal backing to stop any attempt at human reproductive cloning. It recommended monitoring of the outcome of embryo research projects and pointed out the need for guidelines illustrating the meaning of 'serious' disease, as required in the new Regulations made under the HFE Act.

However, there are still many people who feel that it is deeply wrong to use embryos in this way. Although the law in the UK is clear that embryo research performed within carefully regulated parameters is acceptable, we can still ask whether the law has got it right.

One of the ethical arguments in favour of embryo research is that it may provide benefits to adults and children who suffer from serious diseases. That is, embryo research might further the interests of other human beings. But how can we determine *whose* interests are important? Perhaps embryos themselves have interests. One of the greatest problems which besets human beings is the question of how to establish what entities are morally important, and what kinds of acts are harmful to those entities.

These problems are manifest on a broader scale in everyday life. For example, many of us do not raise quibbles about the farming, killing, and eating of animals, or their use in research. Does this mean that harming animals is acceptable? If so, why? More perplexingly, if this is the case, then why should cruelty to animals be a matter of concern, for example the abuse of pets? Why do some people have such strong views against hunting when fishing is relatively uncontroversial? Are these distinctions simply illogical, or is there some sensible underlying explanation?

Three arguments are commonly used to justify killing and performing research on animals.

1. The claim that the entity in question does not suffer.
2. The claim that only some entities' suffering is morally important.
3. The claim that causing harm to some entities is outweighed by other benefits.

When attempting to determine whether performing certain procedures on particular entities is acceptable, we can try to establish whether the beings in question suffer as a result of the procedure. It is suggested that our interpretation of animals' behaviour is unavoidably skewed by anthropomorphism: as humans, we attribute to them the feelings that we would have in such a situation. Likewise, it might be said, if we feel uneasy at the idea of research on embryos, it is simply because we do not like the idea of someone dissecting us. This feeling has perhaps been reinforced by the propensity of the British media to write of embryos as though they are miniature babies suspended in time.

A proponent of the argument that animals do not suffer would suggest that in fact neither the hooked fish, the hunted fox, the battery hen, or the laboratory rat is harmed. Our concern for animals is simply the misguided product of our overactive anthropomorphic imaginations. Of course, there are many assumptions embedded in this type of argument, namely that harm is related to pain, suffering, or consciousness. In fact harm itself is a complex concept which does not necessarily equate easily to any one definition.

If we wanted to prove that animals do not feel pain or suffer, scientific tests could be carried out designed to show that their responses to painful stimuli are in fact merely instinctive, and that they do not experience pain in the way that conscious creatures such as ourselves do. Some years ago, experiments were carried out to demonstrate whether fish can feel pain (the answer, apparently, was yes).[10] The morality of such an experiment might be questioned, but even if we had no moral qualms regarding trying to establish the same 'proof' for an embryo, it would be impossible to do so. A fish, though vastly different from a human being, can move and react. It has a recognisable nervous system, and can behave in ways which we can at least attempt to interpret. The embryo can do none of these. But does this mean that it does not suffer?

Few people argue that embryos consisting of only a few cells are capable of suffering in the way that we understand it.[11] Therefore we might conclude that embryos which are destroyed in the process of research are *not* harmed and that such research is therefore morally acceptable. Similarly, where fertility treatments involve the creation of surplus embryos which may be biopsied, implanted, destroyed, or left unclaimed in clinics, we might claim that—whatever the other moral implications of such behaviour—it is not harmful to the embryos per se.

Loss of potential

However, even if embryos cannot feel pain, or react to painful stimuli in ways we can meaningfully interpret, they might be harmed by the loss of potential involved when they are destroyed or discarded. An embryo which cannot experience or react to suffering is nevertheless a potential human being.

One argument against this might be that a high proportion of naturally-conceived embryos are lost because they fail to implant in the uterus.

Natural wastage of fertilized eggs is exceptionally high in human beings. Not every embryo, whether conceived naturally or deliberately created, has the potential to become a human being. However, there is an important moral difference between what happens naturally and what we undertake to do deliberately. Murder cannot normally be defended on the grounds that that person would inevitably die at some point anyway. An action which is deliberately carried out calls for a more complex analysis.

From this perspective, while it is true that the potential of many embryos is thwarted by the processes of nature, we cannot infer from this that it is right deliberately to destroy embryos. If embryos are harmed by having their potential thwarted, it could still be wrong to bring this about deliberately no matter how often it may happen naturally.

But perhaps not all embryos have equal potential. This is why some people believe that performing research on 'surplus' IVF embryos is better than creating embryos especially for research. If embryos are due to be discarded, it might be claimed that they do not have the potential that an embryo destined for implantation has. However, there is something unsatisfactory about this. After all, the fact that a 'spare' embryo is not going to be implanted and become a child is merely a contingent fact. If it were implanted, it could have just as much potential as any other embryo. In effect, it is the decision not to implant the embryo that robs it of its potential. Thus, if being robbed of potential is harmful, then making the decision not to implant an embryo could be just as bad as actually dissecting it.

Essentially, what appears to emerge here is that while embryos consisting of a few cells may not experience pain in any way which we can recognize, the loss of potential does seem to be an inescapable part of embryo research. Is this loss of potential to be viewed as a harm, and is it significant enough to provide a reason against performing embryo research or fertility treatments which generate 'spare' embryos?

One of the difficulties related to the idea of potential is that it is hard to know where to draw boundaries. For example, a sperm and an egg also have the potential to become an adult human being, and sperm and eggs are often stored in fertility clinics. Should we feel that we are thwarting their potential if we fail to fertilize them? Menstruation, contraception, and even celibacy are also implicated in a loss of potential since eggs that could otherwise be fertilized are lost. Loss of potential alone does not seem to give a clear-cut answer to why embryo research might be wrong, since it applies to other practices which—intuitively, at any rate—do not seem to raise ethical concerns at the same level.

Conversely, if we agree that loss of potential does not constitute harm to embryos used in research, this may put pressure on other intuitive ethical convictions, such as believing that it is wrong to kill small children. Suppose, for example, that it were possible to perform research on babies whose parents did not want to raise them. Such research could be performed under powerful anaesthetics, and need not necessarily cause any pain or obvious suffering to the subjects, who would be humanely terminated at the end of the experiment. Most of us—including those who have no objection to embryo research—would feel that this would be utterly abhorrent. But what is the difference?

In both cases, no suffering is involved, and in both cases there is a loss of potential. Is there any reason to protect unwanted babies but not unwanted embryos? Perhaps the simple answer is the age-old conviction that there is a moral boundary—wherever it may lie—between an unborn entity and a baby. At any rate, it does not seem that a consideration of potential alone can provide a conclusive answer to how we should treat embryos outside the body.

'Personhood' and special attributes

Another approach to ascertaining the moral interests of entities rests in the idea that being a human person is morally special. Harm to a 'non-person' may be acceptable, whereas harm to a person is not. Again, this is an apparently simple argument which yields a wealth of complexity on further investigation. Crucially, we need to know how to define 'personhood'. The term may be taken to imply a being who possesses one or more morally-significant attributes, such as consciousness, or perhaps the ability to reason, or to communicate, or to value its own life. Clearly, on these cognitive or intellectual criteria, the fourteen-day old embryo fails on all counts, although again if the potential to have these capacities is regarded as being important, this position would lead back to the argument previously discussed.

The belief that personhood relies on the possession of certain capacities poses some interesting challenges. How we should treat human beings who lack those functions? If self-consciousness is the fundamental component of personhood, for example, does this mean that experiments can be performed on someone who is comatose? Would it be acceptable deliberately to create people who lack these faculties, in order to perform research on them? For most of us, the answer is emphatically no. People must be treated as having moral worth regardless of their capabilities. But we cannot reasonably claim that there is one particular attribute which is sufficient for personhood if we still insist that people who lack that capacity are to be treated as having moral worth.

For many this might be the end of the issue. However, some people, among them the philosopher Peter Singer, have pushed logical consistency to its moral limits. Singer suggests that cognitive ability is the essence of personhood. Because of this, he maintains that human beings who lack these faculties do not have full moral worth. He goes so far as to suggest that where the higher apes' cognitive faculties exceed those of particular human beings, it would make sense to regard those apes as having more moral worth, or as being persons. Thus, he opposes research on the higher primates, but is willing to countenance infanticide.

This is an unpalatable conclusion for many people. Yet similar problems arise whatever attribute is fixed on as the essence of personhood as, invariably, not all human beings share all the attributes that are thought to be important. One way to avoid this difficulty is to re-construe the idea of personhood. Instead of being associated with particular faculties, we might place value simply in being identifiably human. This would mean that all humans, regardless of their abilities, are entitled to special consideration. People in comas, newborn babies, and all other human beings, and also embryos would presumably fall within this category—embryos are not, after all, members of any *other* species—and those who take this view would accordingly disapprove of embryo research.

This approach also circumvents one of the problems involved in the argument from potential. Sperm and eggs have the potential to become human beings, but are not 'human' in the way that embryos are, since they do not possess the full genetic complement of an individual human being. Thus, if the essence of personhood is to be located in being human, this might enable us to make a coherent moral difference between eggs and embryos. It would also support a distinction between humans and other animals, however advanced their cognitive faculties may be.

However, can we really consider embryos as being 'persons' like us who deserve equal consideration? Peter Singer's views on the acceptability of infanticide can be considered to be extreme, but is it really possible to view the death of an eight-cell embryo as being a tragedy on the same scale as the death of a fully fledged human being? Given the choice between killing an embryo and an adult human being, or even a cat or dog, which would we find easier?

An embryo consisting of eight undifferentiated cells is different from an adult human being in almost every way imaginable. The embryo cannot perceive, it has no heart or brain, and it is entirely dependent on other people for its existence. As is frequently reiterated, an embryo of up to 14 days old is simply a cluster of cells. Yet we are all, adults and embryos alike, composed of clusters of cells. An adult human being consists of rather more cells than an embryo, but this fact hardly seems a good reason for moral differentiation. If it were, then it would seem to follow that larger people, who are composed of more cells, should receive preferential treatment.

Of course, those who regard embryos as merely tiny clusters of cells are not really suggesting that the more cells a person has, the more moral worth that person will have. In fact, the number of cells argument can be regarded as something of a red herring. If there is a moral difference between an embryo and an adult human being, it is not directly related to size, or the number of cells. If someone were shrunk, or lost their recognisably human shape, surely we would not suddenly be justified in performing experiments on them. So it is not simply the embryo's size or shape which is morally relevant here and again the arguments seem inconclusive.

If there is a moral difference between embryos and born human beings, it might lie in a complex relationship between some or all of the factors discussed above. Size, possession of certain attributes, being recognizably a human being... in isolation, none of these factors seem conclusive. For some people the discussion in this chapter may sway them in favour of embryo research; others may interpret the reasoning to weigh against it.

Social harms

There is another consideration which should also be addressed here. Even if it were demonstrable, first, that embryos cannot be harmed, and, secondly, that significant therapeutic benefits might be gained from embryo research, such research might be regarded as a kind of symbolic harm to society. If this is the way we are willing to treat the most vulnerable and helpless members of the human species, what does it say about us as moral beings?

It is interesting to observe in this context that some of the states that are most protective of the embryo are willing to accept other practices which we might regard as being unethical, such as capital punishment. Conversely, those states with few prohibitions about the treatment of the embryo are the ones most likely to have protective human rights laws (the UK, Scandinavia, Australia, Japan and Canada, for example). Seemingly, then, whatever social or symbolic moral value may appertain to embryos, respect for them does not necessarily translate into a respect for humanity in general. Likewise, embryo research need not inevitably lead to social disintegration.

However, this will depend to some extent on the society in question and the scale of opposition. As mentioned previously, in Britain the legal status of the embryo and of embryo research were settled in the HFE Act 1990, which was based on the findings of the Warnock Report. This report concluded, after weighing the arguments on both sides, that the embryo 'deserved respect'. Respect for embryos was felt not to preclude their use in research, or the deliberate creation of embryos specifically for research. However, it was stipulated that embryos should only be used in research designed to tackle 'serious' disease, and they were not to be used frivolously. It may be that this respect for the embryo was an attempt to forestall the possible social harms that might have been associated with more unrestricted approach to embryo research.

Respect for the embryo and 'non-frivolity' are somewhat vague concepts, but these concerns are at the bottom of the painstaking care taken in the oversight of embryo research. All research that involves the creation, storage, or use of human embryos outside the body must be licensed by the HFEA. Licences cannot be granted unless the research is deemed to be necessary or desirable for the purposes outlined in the HFE Act. All research applications submitted to the HFEA must also have Research Ethics Committee approval. The project must explain its objectives, protocols and why the research is necessary.

The HFEA holds data on the number of embryos created for IVF at each clinic, and also the number of embryos used in research, so that this ratio can be monitored carefully. Peer reviewers are routinely asked to comment on the numbers of embryos that the clinic proposes to use in the project. In the 1990s, 118 embryos were created solely for research purposes, but 48,000[12] were donated for research as surplus to IVF requirements.

If embryos are not harmed by having research performed on them, why the insistence on non-frivolous use, and the efforts spent on monitoring this research? Does this imply that embryos would be harmed if used frivolously, or if treated with a lack of respect? If not, why make these stipulations?

Perhaps respect for the embryo is not necessarily designed to protect embryos as such, but partly to appease those who feel that this research is wrong. Where views are as polarized as they are in the context of embryo research, appeasing measures may afford little comfort to those who think that a grave moral wrong is being perpetrated. However, to those who are uncertain, perhaps these measures provide some reassurance.

Given the difficulty of establishing what constitutes harm to the embryo, and the questions over what we may achieve through embryo research, it is not necessarily ridiculous to attempt to mitigate any harm involved in embryo research by ensuring non-frivolous and respectful use of embryos. It is part of an admission that we do not fully comprehend the moral dilemmas involved, and an effort to ensure that if harm is involved, it is minimal, and that the risk is not undertaken lightly.

Technology has opened up before us some uncertain moral territory. For scientists themselves, respectful treatment of embryos may be an appropriately cautious and thoughtful way of conducting their research. Embryos are hard won; they are the products of the human body; they represent difficult choices and sacrifice for those who have taken the decision to donate them or the gametes from which they are made.

None of these arguments can be seen as conclusive: they are not meant to be so. However, in a field where dispute is inevitable, and where moral uncertainties abound, it is important that these difficulties be admitted, and that the reasoning behind decisions concerning embryo treatment should be honestly addressed.

Obtaining eggs

Any advances in embryonic stem cell research clearly rely on a supply of embryos, or at least of eggs and sperm from which to create embryos for research. However, donated gametes (and embryos) are in extremely short supply. It is worth considering also that if fertility treatments become more successful and less dependent on the creation of surplus embryos, this source would, in effect, dry up. Now, and in the future, therefore, it seems likely that men and women may be called on to donate their gametes in order to facilitate embryo research.

In the UK the availability of donated sperm is extremely limited, despite the fact that obtaining it is not invasive or dangerous to the health of the donor. The harvesting of eggs on the other hand, is both invasive and risky. Women who donate eggs undergo a gruelling regimen of drugs, followed by a surgically invasive retrieval procedure. These interventions are not risk-free, and women can die from the side-effects of the drugs involved. Unsurprisingly, donated eggs are also very scarce.

The demand for eggs places a strain on the idea that our bodies and our gametes are somehow sacrosanct and should have no market value. The truth is that the demands of scientific research as well as the rise of fertility treatments have generated a flourishing market in these commodities.

Case 32: Woo Suk Hwang

On Monday 8 August 2005, the BBC News website featured an article[13] on one of the stars of cloning and embryo research. Woo Suk Hwang, a Korean scientist, was revered throughout the scientific world for his ability to achieve astounding results where others had failed.

The news story described Hwang's success obtaining stem cells from embryos cloned from the cells of adults. Hwang was also renowned for his success rate at cloning animals. A self-described workaholic, Hwang represented his achievements as a way of benefiting mankind. While acknowledging that embryonic stem cell work was opposed by some people, he cited the strict ethical and regulatory guidelines under which his research was conducted.

However, in November 2005, reports emerged that Hwang had broken ethical protocols by using eggs obtained from his own researchers. Later in the year, claims were made that Hwang's cloning and stem cell successes had been falsified. An academic panel investigated the allegations, and it was confirmed that his work had been fabricated. Formerly a national idol, Hwang was now in disgrace.

The rise and fall of Woo Suk Hwang, apart from constituting an interesting lesson on the dangers of hype, demonstrates the ethical difficulties which may ensnare scientists. The use of eggs obtained from colleagues is generally agreed to be unacceptable in international protocols. This is largely due to the possibility of coercion (for example, a young woman is offered a job on the understanding that she will be willing to donate her own eggs). However, the concern which surrounded this news seemed a little overblown in some respects. It was obvious that Hwang was obtaining large numbers of eggs and some observers were perhaps unsurprised when they heard the news of his unethical conduct in this respect.

The uncomfortable truth is that reproductive technology and embryo research have between them created a market in those hitherto most sacrosanct of tissues: sperm, eggs, and embryos. Much agonizing takes place over whether we should allow our bodies to be treated as commodities, and these debates frequently erupt in the arena of organ donation. The UK prides itself on the dissociation between finance and healthcare. Payments for bodily commodities are not countenanced, and such practices are regarded as being the province of countries.

This squeamishness in relation to money and biological commodities may in itself bring about some ethically dubious results. Some British fertility clinics openly advertise services in facilitating egg procurement from other countries. The HFEA imposes strict conditions on the import of eggs. However, travel abroad for treatment with donated eggs cannot be controlled.

On a more general level, it has been suggested that encouraging women to donate eggs for research at all is simply unethical because the therapeutic benefits from stem cell research may be so distant.[14]

Do the benefits outweigh the harm?

Acceptance of embryo research is based on an assumption that it will bring benefits which outweigh whatever harms may be involved. This raises another important question: will these benefits really arise? Certainly fertility treatments and the understanding of early human development have improved as a result of embryology research, and these benefits are passed on to subsequent patients. However, one of the favoured arguments in support of embryo research focuses on the possibility of developing embryonic stem cells which could be used to cure hitherto intractable diseases such as Parkinson's, Alzheimer's, some cancers, diabetes, motor neurone disease, and many others. But some commentators have suggested that the supposed miracle cures we expect from stem cell research may never materialize, or will be far more limited than we have been led to believe.

This is a question which is difficult to answer: to an extent no scientific research is guaranteed to bring benefits. We cannot be certain how things will turn out, because if we *were* certain, we would not, of course, need to do the research at all. Research implies an experimental engagement with the unknown. However, if there is reason to suppose that there could be significant potential benefits, it could be that these would outweigh some of the difficulties related to embryo research. Even if there were little evidence of immediate benefit, it might be suggested that there are general benefits to mankind in the quest for more effective medicines.

One of the peculiar attributes of humankind is its instinctive urge to push forward at the boundaries of knowledge. For some, this may in itself seem to be a commendable goal. This general movement toward greater knowledge and technological achievement will necessarily involve some failures, and some explorations of blind alleys, but it cannot be known in advance which avenues will prove fruitful, and which will not.

At the very least, if there are benefits which could be gained from embryo research, they cannot be realized unless embryo research is allowed to proceed. And we cannot find out the answer to the question as to whether the benefits will materialize without going ahead in order to find out. Of course, if the human endeavour towards greater knowledge and scientific power were not regarded as being an intrinsically ethical aim, then embryo research would not be justified. However, this is a position which applies to the aims of research in general, and is not exclusively about the ethics of embryo research.

Global concerns

Even if benefits are gained from embryonic stem cell research, the kinds of disease for which they might provide treatments are largely late onset problems that mainly affect affluent Western nations. In some countries, other medical, social, and economic factors mean that life expectancy is so low that people would be statistically unlikely to suffer from Parkinson's or Alzheimer's disease, as they would probably already have died from other causes before these became an issue. Can it be reasonable to pursue solutions for these diseases when the need for more basic healthcare is so pressing in some parts of the world?

There is certainly something uncomfortable about the disparities in wealth and healthcare provision around the world. However preventing stem cell research is unlikely in itself to remedy global poverty or low life expectancy. It would be cruel, too, to dismiss diseases as being of less consequence because they are 'late onset'. Knowing that one is prone to a fatal genetic disease may be a dreadful burden to bear, especially if there is added to it the possibility that more might have been done to prevent this situation arising. 'Late onset' refers to the fact that the disease may not manifest itself fully until patients are in their 30s or 40s—the very stage of life during which we are most likely to have the heaviest responsibilities for a young family and a career. Very few of us will achieve our life's work before that age, unlike the rare case of Mozart!

Moreover, if we believe that there is a 'trickle-down' effect, we might justify the research to ourselves on the grounds that it may provide some benefit to some disadvantaged people at some stage in the future. If our research serves to bring into existence new techniques and cures, there is no reason to suppose that when economic and social difficulties are resolved they should not benefit humanity on the broadest possible scale. It is worth remembering that research undertaken in the twentieth century, such as that which led to the discovery of penicillin or treatments for blindness, has been of enormous value all over the world.

These are not questions which apply uniquely to stem cell research. It is unclear how much of our resources should be diverted towards the needs of other countries and communities, and there is no easy answer to this. However, the fact that such a gulf exists and that current research preoccupations are only likely to increase this disparity, this may well prove to be a cause for concern that does not necessarily revolve around the well-rehearsed issues related to the status of the embryo.

Broadening the scope

Many of the prospective benefits discussed so far have been related to the possibility of therapeutic cloning. There may be benefits associated with embryonic stem cell research which do not rely on this uncertain and technologically intricate procedure. One possibility is the development of a number of stem cell lines not matched to individual tissue types, but banked and made available to patients in general. Thus when a patient needed stem cell therapy, the closest available match from the stem cell bank could be chosen. This kind of approach would bring down the costs associated with stem cell therapies, as well as perhaps making them available to a far wider community. However, there would obviously be drawbacks here: since these stem cell lines would not be individually matched, some degree of rejection and a corresponding need for immuno-suppressants might become necessary.

Legal challenges

In Britain, work on stem cells derived from embryos left over from fertility treatments is legal, provided that the embryos have been donated for research purposes with the informed consent of both parents. It is also legal to create embryos specifically for research where this is deemed to be necessary. Research involving embryos must be approved by scientific peers as well as by ethics committees. The licence from the HFEA takes into account quality approval, justification, legal compliance, consent, and yearly accounting. This is strict but effective and gives researchers an arena of safety and respectability in which to proceed secure in the knowledge that they are operating within the regulatory framework.

Licences are not simply granted as a matter of course to any person who wishes to perform research on embryos. The HFEA needs to be assured that the research could not be carried out by other means. The research must not be undertaken from mere curiosity, but must be necessary or desirable for the treatment of infertility or serious disease. 'Seriousness' is, as discussed elsewhere, a fluid concept, changing according to culture and context. Over time, tolerance of ill-health may diminish and the set of diseases regarded as serious will expand. However, maintaining this concept of seriousness even while accepting that it is flexible, ensures that embryos are not used frivolously, and that the scientific goals are clear and justifiable.

Under British regulation a sample of every stem cell line whether derived from a cell taken from an embryo or adult must be deposited in the UK Stem Cell Bank, which was set up in 2004 by the Medical Research Council and the Biotechnology and Biological Sciences Research Council. Researchers will only receive a licence if their project cannot be undertaken using existing stem cell lines in the Bank. This ensures that the numbers of embryos used in research are kept to a minimum and that research is not duplicated.

However, these provisions have come under attack from the Pro-Life Alliance, an organization devoted to the protection of embryos and foetuses. The embryo is defined in section 1 of the HFE Act as a 'live human embryo where fertilization is complete'. This definition was challenged in court when the UK became the first country in the world to legalize and regulate therapeutic cloning or cell nuclear replacement.[15] The Pro-Life Alliance argued that there could be no 'embryo' without 'fertilization' and that since the procedure of cell nuclear replacement stimulates an embryo into growth by artificial means and not by the penetration of the egg by sperm, this did not constitute an embryo.

This argument succeeded at first instance in the High Court. Its success meant that the HFEA would have had no control over cell nuclear replacement research and that there would have been no means of stopping human reproductive cloning. The government appealed against this judgment and won a reversal in the Court of Appeal, where a constructive approach was adopted to the intentions of the HFE Act in banning cloning. The Pro-Life Alliance's appeal to the House of Lords was rejected. This decision on the one hand upheld the democratic or parliamentary approach to this new field in that it supported the remit of the existing law, but it also enabled the regulators to extend their reach.

This may seem an odd case for the Pro-Life Alliance to have brought. Its remit is, after all, to protect embryos and prevent their destruction. This was unlikely to be furthered by demonstrating that some embryos fall outwith the regulator's jurisdiction. However, there was a certain logic in the Pro-Life argument, in that—arguably—it exposed loopholes in the legislation that would allow for unethical practices. The challenge certainly served to demonstrate that the existing controls were weak and needed to be strengthened by the government.

The government responded to the challenge with a speedily enacted new law, the Human Reproductive Cloning Act 2001, which provides that 'A person who places in a woman a human embryo which has been created other than by fertilisation is guilty of an offence'. This could be punished by up to ten years' imprisonment. This statute should be sufficient to stop human reproductive cloning in the UK unless the day comes when embryos can be brought from fertilization to the birth of a baby by use only of an artificial environment and not the uterus.

In 2002 the Lawyers' Christian Fellowship challenged the legality of two licences granted by the HFEA. The licences had been issued under the new wider categories of research permitted by the 2001 Regulations. It was claimed that the research was focused on diseases which were not 'serious', and therefore failed to meet the qualification imposed by the Regulations. Examples have been given of diseases likely to be the subject of stem cell research including: repairs for spinal cord injuries, replacing lost heart muscles in cases of congestive heart failure, bone cells in osteoporosis and liver cells in cases of hepatitis, nerve cells in Parkinson's disease and following stroke, and replacing insulin producing cells in diabetes. Eventually this case was withdrawn, presumably on the grounds that the point had been made even if the case itself had not been won.

Other ways of obtaining stem cells

Given that even in the UK where embryo research is legal and well regulated, the ethical issues around embryonic stem cell research are still extremely fraught, it is relevant to seek ways of performing research which do not require the destruction of embryos. A number of possibilities have been proposed; however, on closer inspection many of them seem to raise as many problems as they purport to solve.

Single cell biopsy

We know from PGD that a cell can be removed from an embryo without necessarily harming or destroying that embryo. In PGD this is done in order to establish whether the embryo is carrying a particular genetic flaw. However, it could equally be performed simply in order to obtain a cell for research.[16] This might be a way to circumvent the US federal funding ban. A single cell detached from an early embryo can be grown in culture to create stem cell lines, and the embryo itself could then be implanted into a woman as normal. (This would be contrary to current

UK law.) Keeping the embryo alive would address the ethical concerns of many, but it is not as efficient as culturing the entire eight-cell embryo. However, it is perhaps questionable under what circumstances a woman will want to carry an embryo which has been biopsied in this way. It seems understandable that parents at risk of genetic diseases may accept biopsy as a way of avoiding a terrible affliction for their child, but to impose the risk—even if small—of performing a biopsy on an embryo with no intended benefit either to child or parent might seem to be unappealing to prospective parents.

Even if the removal of cells for research was carried out in conjunction with PGD, removing more cells than would be necessary for the clinical task in question would seem to raise difficult ethical issues.

Dead embryos

In recent years, British scientists have taken cells from dead embryos and grown them the stage where stem cells might be extracted.[17] The scientists used embryos that had died naturally during IVF treatment. The embryos in this experiment had stopped developing a few days after fertilization. Nevertheless, the question was immediately raised of how one would know that the embryos were dead, and whether there was something wrong with those embryos that caused the arrest of their development. Other ethicists however regarded this new process as akin to organ donation from dead patients, with no more concerns than surround that process.

Parthenogenesis

The properties of cells in culture are sometimes surprisingly different from their properties *in vivo*. One of the most fascinating capabilities of egg cells may yield an opportunity for future embryo research without raising the ethical and political difficulties involved with the use of specially created or leftover IVF embryos. When eggs are kept in culture, they sometimes start dividing as though they had been fertilized. It is not altogether clear how or why this happens, but it is a well-documented occurrence in many species, including human beings. Could we deliberately bring about parthenogenesis in eggs, and thus use these 'embryos' to derive stem cell lines? The fact that this has been achieved in rabbits might imply that it should at least theoretically be possible in human beings.[18,19]

The question remains why using this type of embryo should be ethically preferable to using any other? One possible answer might be that parthenogenic embryos are not viable: they do not have the capacity to become fully fledged human adults. For this reason, performing research upon them would not seem to thwart their potential. However, some people may nevertheless be opposed to the deliberate creation of an entity with the sole purpose of ending its 'life' in order to perform research. That is to say, only those who object to embryo research on the grounds of an embryo's potential for life would be appeased by the use of parthenotes instead. Some people feel that the moral wrong in embryo research is a deeper and more symbolic matter: it does not centre so much around what the embryo is, as the willingness of the scientist to create and destroy entities for the sake of research.

Finally, even in the context of viability, research on parthenotes might be less of a perfect solution than one would think. After all, it is the action of the scientist who deliberately brings about parthenogenesis that effectively removes the unfertilized egg's potential for viability. Prior to this, the egg has as much capability of being fertilized and forming a viable embryo as any other egg. Some of the ethical weight involved in this question is surely tied up in this choice which results in a loss of potential.

Of course, arguments related to potential are tricky, as they can seemingly go on ad infinitum (eg if an unfertilized egg has potential, then so do sperm, so are we thwarting millions of potentially viable individuals if we fail to ensure that each sperm fertilizes an egg?) However, the problem remains that for many people viability is not the central issue in the moral wrongness of embryo research. It is also important to note here that the fact that parthenotes are not viable may in itself be a hindrance to research: genetic anomalies in the stem cell lines derived from them may be a possibility, in which case they will offer no solution to the research problem.

Finally, there is a problem which arises with reference to all the alternative methods of deriving stem cells discussed here. That is, the whole point of interest in stem cells turns on their capacity to differentiate into other cells: on their potential, in fact. If stem cells have the potential to form a viable embryo, even if they were not derived from a viable embryo this could pose problems for those who disagree with embryo research.[20] This is an interesting issue, especially as it could in theory apply to stem cell lines derived from adults as well as embryos.

The role of the HFEA

Issues related to stem cell research are often discussed as if they could be decided purely by ethics, law, and science. In reality, the realms of ART and embryo research are battlegrounds fought over by legislators jealous of their power, desperate patients, clinicians and companies with considerable earnings, and religious pressure groups. The only way to keep the peace is by comprehensive regulation by as neutral and expert a body of people as can be assembled. Any other solutions result in distortions and inconsistencies.

The political and financial aspects of embryo research regulation can be aligned with various ethical and religious approaches. Should unlimited access to embryos for research be allowed? Or should any such research be outlawed on the grounds that it is unethical to experiment on human beings? The Warnock Committee reached a typically British compromise to this dilemma: in UK law the embryo has been given special status but not absolute protection.[21] In the UK, the embryo is treated as an entity deserving due attention, which means that is to be used only if there is no alternative, its use must be ruled by informed consent of the donors, there are restrictions on exporting embryos, or mixing them with non-human material, and rigorous records must be kept to ensure that every single embryo in research is accounted for.

The position in the UK now is that a wide range of embryo research is permitted under licence from the HFEA. The extended list of purposes for which research involving embryos may be permitted is: promoting advances in the treatment of infertility; increasing knowledge about the causes of congenital disease and the causes of miscarriage; developing more effective techniques for contraception; developing methods for detecting the presence of gene or chromosome abnormalities before implantation; expanding knowledge about the development of embryos, about serious disease or enabling such knowledge to be applied in developing treatment for serious disease.[22]

National differences in legislation/regulation

A. Germany

In Germany, embryo research is illegal.[23] It has been suggested that this conflicts with the constitutional guarantee of freedom of research which also exists in Germany. However, guaranteed freedom of research must clearly encompass some limits. Unlimited vivisection on adult human beings, for example, could yield immensely beneficial results when compared with animal or embryo testing. However, no-one seriously suggests either that this should take place, or that the failure to permit it in law should be regarded as an infringement of scientific freedom. The question, then, is where the boundaries should be drawn, and what should follow from this.

The UK permits embryo research not simply as a result of objective logical and scientific analysis about the status of the embryo (and it is arguable how much this could prove, in any case), but because its cultural and political context has enabled it to do so. Germany's history—perhaps understandably—makes it reluctant to engage on research which is still viewed by many as being experimentation on human beings.

Following on from this, Germany also permits *in vitro* fertilization of eggs only to establish a pregnancy, and the deliberate creation of surplus embryos is not allowed. Thus, the scope for creating 'spare' embryos which, it might be argued, would be wasted if not donated for research, is obviated. This has the additional effect of preventing PGD, since all embryos have to be created with the intention of implanting them. In these respects again there is a wide divergence between Germany and the UK. But despite these stringent prohibitions, stem cell lines can legally be imported into Germany provided that they were derived before 2002.[24] In that year German law proscribed the derivation of new stem cell lines. It has been suggested that there is a certain amount of ethical hypocrisy here. German scientists are permitted to work with the—by implication—unethical products of the research of other countries.

Can it be consistent for a country to outlaw embryo research, and yet allow stem cell lines created by other countries to be imported and also to permit IVF? There seems to be something almost parasitical about this. However, it should perhaps be regarded as a pragmatic move on the part of German legislators. The possibilities of embryo research, as we have discussed, have thrown existing ethical norms and principles into disarray; it is not clear how or where to accommodate early embryos outside a woman's body in the scale of entities which have moral value and/ or legal protection. This being the case, we might interpret the German position as being—like that of the UK—agnostic: there is no objective way of determining whether embryos should or should not have full moral status. Germany has acted on its agnostic position otherwise to the UK. However, its stance is perhaps no more illogical than our own, which permits embryo research, yet strictly controls it and speaks in law of 'respect' and 'protection' of the embryo. The difficulty of attaining consistent and logical approaches is perhaps an inevitable consequence of having our moral systems invaded by new entities with uncertain status.

However, there is another angle to consider here. At present, these questions about consistency or hypocrisy are relatively academic since there are few, if any, treatments available from embryonic stem cells. However, clearly, the aim of research is to develop such treatments. It may be one thing to permit research on ESC lines developed before 2002, but if and when ESC therapies become available, how will Germany and other countries respond? ESC lines developed pre-2002 have all been created using early techniques incorporating tissue from other animals. It is widely held that the risk involved in using any of these lines for the treatment of human beings would be unacceptable. Because of this, ESC lines which have any possibility of being used in therapies would need to be created on 'pure' human feeder cells. Such lines will necessarily have been developed post-2002.

This would put countries such as Germany in a difficult position. For consistency, it would seem appropriate that countries that ban research on these ESC lines should also refuse to allow their citizens access to the treatments which are developed from this research. Yet it is clear that this could have serious consequences, in creating a division between the health outcomes of the different nations, and even perhaps in generating social uprisings. A German citizen whose government forbids a treatment which would be available to a British citizen might well feel aggrieved (albeit that European legislation frees him or her to travel anywhere else in Europe to receive it). This kind of situation would be likely to put democratic countries under strain, since it is unlikely that all of the voters would necessarily have agreed with the veto in the first place. In fact, it is hard to imagine that this kind of restriction would be desirable or enforceable. Rather, the inconsistency of approach would simply be expanded.

This kind of speculation about the future seems to indicate that in fact a determination to achieve consistency in moral reasoning is perhaps neither desirable nor practicable for countries such as Germany—or indeed the UK.

B. *Italy*

Italy was until recently the most unregulated country in Europe. It was renowned for offering fertility treatment to women over 60, as well as providing treatments regarded as ethically dubious in other countries, such as sex selection and embryo splitting. However in 2004 a comprehensive law was introduced.[25] No scientific research on embryos is permitted; the use of PGD is also illegal, as are gamete donation and egg freezing. The number of embryos that may be transferred per IVF cycle has been restricted to three. IVF is available only to heterosexual couples who are married or living together. No posthumous treatment is allowed. In effect, the law introduces a set of prohibitions rather than constructing a general regulatory framework for the conduct of assisted reproduction and/or research.

The Italian approach penetrates a long way into the area of medical discretion in, for example, banning embryo freezing and limiting the number of embryos that may be transferred in one cycle, thereby reducing the success rate. The law goes even further than the report of the Dulbecco Commission that preceded it, which had recommended that stem cell research should be allowed on surplus embryos. The Italian law has been condemned by some European research organizations since it greatly reduces the chances of an infertile woman bearing a baby. This may be regarded as a problem by some: is respect for the embryo being elevated at the expense of women who undergo fertility treatment? However, the electorate was given the opportunity to vote against this law, and did not do so.

C. *The USA*

The position in the US is perhaps the most troubling and inconsistent of all, and the results of this affect the whole world. The stance taken stems from a combination of politics, business, religion, and an aversion to federal control. Any general regulation or ethics directive is seen as undermining the doctor–patient relationship and imposing bureaucracy on medicine. Lack of overarching regulation has led to rule by market forces, a general free-for-all which in turn leads to doctor/patient/baby conflicts of interest, abuses, and dubious genetic 'cures'. The regulation that has been attempted by professional bodies in the US has proved ineffective because no consensus has been reached on the main issues and the guidelines are unenforceable at law.

These issues are all bound up with the very sensitive American position on abortion and the tensions over childbearing issues between religious forces on the one hand and the constitutional rights to privacy and liberty on the other.[26] Any legislation that interfered with for example the 'right' to clone could be open to constitutional review by the courts. Much as the US needs federal regulation of embryo research, it is particularly difficult within the US because of the Constitution, guarantees of State and personal autonomy, and the political/religious lobby.

American law prohibits the use of federal funds for the creation of human embryos for research purposes. This means that private clinics and company laboratories are free to undertake research as long as they can pay for it themselves. In 2001 President Bush announced that federal funding may be used for research on existing stem cell lines, but no others. At this time, the Council on Bioethics was formed, under the chairmanship of Leon Kass, a conservative thinker. The effect of the law is that no live embryo may be destroyed for research purposes.[27] Federal funds may be spent on adult stem cells, to which no constraints apply.

There is little open and informed debate of the issues in the US because there is no public accountability for embryonic research and no unified voice speaking for it, only presidential diktat. Private companies perform much unsupervised research, which, while profitable to those involved, may be hard to square with concerns over public health. The current position seems undesirable on many levels, at least to the British point of view: the public sector is obstructed while private work is unregulated and driven by commercial concerns.

Recently, however, some States, led by New Jersey and California, have legislated for State funding for stem cell research. California Proposition 71 of 2004[28] granted $3 billion funding to stem cell research. The measure included the establishment of the California Institute for Regenerative Medicine to regulate and oversee stem cell research. This is likely to lead to a movement of scientists from restrictive States to those that are more liberal, just as the more permissive regulatory environments of other countries has attracted American researchers abroad.

Interestingly, the Act has a preamble stating that half of the families in California have a member who has or will suffer from a medical condition that could be treated with stem cell therapies. The focus, then, is very much on foreseeable benefits. It is perhaps questionable how long these benefits might take to materialize. The Act is designed to plug the gap in federal funding and to shift the emphasis of health care towards prevention rather than expensive cures. It will presumably give a boost to the prestigious Californian universities' research programmes. Human reproductive cloning remains forbidden, but otherwise all types of research on all types of embryos may be carried out.

Accountability is to be achieved by open meetings and annual reports to the public. The Independent Citizens' Oversight Committee of twenty-nine members governs the California Institute for Regenerative Medicine and will include patient, university, and research representatives. The Act directs the Committee to establish standards concerning informed consent, controls on human research, the prohibition of compensation to donors, privacy laws, and time limits for obtaining cells (eight to twelve days after fertilization). There are similar laws in New Jersey[29] as well as a Stem Cell Institute of New Jersey, with a grant of State funds, albeit which is very small compared to California.

It seems that there will soon be a patchwork of regulation across the US and a scattering of States where funding for stem cell research will be provided. Inevitably, this will lead to scientific tourism, a reflection of the world situation. This may be seen as a further argument in favour of national control and a national regulatory body. The President's Council on Bioethics cannot fulfil this function since it is allegedly composed of members chosen for their conservative views appointed by the President, and it has no real powers.

How do the various approaches compare?

What does this brief comparison tell us about the ideal regulatory framework? There may be comprehensive regulation or private rights and prohibitions, such as in Italy. There may be regulation or a free market as in the US. There may be regulation by quango or by legislators, which is the situation in the UK. An analysis of the various national structures of regulation has been made by DG Jones and CR Towns,[30] who describe four types of regulation of stem cell research.

The first category prohibits all human embryo research—Ireland, Austria, Norway, Poland. The second grants permission to use stem cell lines already in existence before a certain date—US, Germany. The third permits research only on embryos surplus to IVF—Canada, Greece, Finland, Hungary, Netherlands, Taiwan, Australia. The fourth also allows the creation of embryos specifically for research—UK, Belgium, Israel, Singapore, Japan, South Korea, Sweden.

If permission is given to use stem cells created before a certain date (the second category), they may be inappropriate for research, as clinical conditions and requirements change. Eventually the stem cell lines will be too old. Allowing only research on existing lines therefore limits the possibilities and reduces the likelihood that the putative benefits of stem cell therapies will be achieved. It is also worth noting that many of the countries that prohibit the creation of new stem cell lines on the basis of the sanctity of the embryo allow the destruction and creation of embryos for IVF. It is questionable whether this is based on a presumption that there is a moral difference between the two endeavours (ie research and reproduction), or is simply a failure to apply the same consistency of reasoning in both cases.

The use of surplus embryos (the third category) makes a utilitarian use of embryos that are in existence and would otherwise perish. This approach is premised on the idea that to allow such embryos to be discarded would be wasteful. Again, countries that adopt this approach frequently allow the creation and destruction of embryos for the purposes of IVF.

The fourth category, the most liberal one, maximizes research opportunities and secures a diversity of embryos for research, including ones that will be compatible for the purposes of transplantation.

So how can we make sense of this diversity of regulatory approaches? Is there a right answer? Another way of categorizing the different national attitudes to embryo research is to look at the contenders, the winners and losers. When no research is allowed at all or only on old stem cell lines, this seems to uphold the intrinsic value of the embryo, but scientists, businesses, and prospective patients are disadvantaged relative to those of more permissive states. However, if the prohibition can be evaded by import or export of gametes, or research tourism, then commerce benefits but patients and domestic researchers are disadvantaged. If research is allowed only on surplus embryos, the likelihood of benefits for future patients may be reduced. If the most liberal attitude is taken, patients, scientists, and businesses may benefit, although the opponents of research may claim that embryos themselves are devalued, and perhaps that society itself is morally damaged.

The questions which this raises are interesting. Whose benefit should we be taking into account here? If we adopt the first approach, seemingly the embryos are the only beneficiaries. If the most liberal approach is taken, arguably all parties except the embryo benefit. Clearly, much of this issue turns on whether embryos can be benefited at all, or whether it is in their interests not to have research performed upon them. However, there may be points to make about the other benefits involved here too. We assume that patients may be the prospective beneficiaries of stem cell research, and indeed it is this hope that propels the research forward. However, there are broader questions to be asked here about the nature of the benefits that may accrue, as well as the timescales involved, and the resources required. Will embryonic stem cell research, for example, benefit patients more than if resources were redirected to some other, less controversial, area of research?

This is a difficult question to answer, but it is an important consideration, which sometimes becomes swamped in the more emotive concerns. It is perhaps not the job of the regulator to make these broader judgements. Nevertheless it is important to remember that embryonic stem cell research does not guarantee a flood of cheap and accessible cures to all the ills which afflict human beings. Part of the result of the polarization of the debate described above is that occasionally claims are made which exceed what can reasonably be expected. If scientists press forward on the basis that medical miracles are just around the corner, the credulity of the public may be stretched too far. But if research is held back in order to fund more immediately realisable objectives, scientific endeavour is thwarted and advances that could possibly benefit everyone will not come about. These arguments are, of course, not specific to embryo research.

In essence, the problem of the status of the embryo, and the difficulties associated with religious, cultural, and political variations and their effects on regulation, seem insoluble. Perhaps the best that we can hope for is that in the UK our regulatory system is coherent and flexible to public opinion, and that we do not oversell the purported benefits of stem cell therapies. We need to recognize that our approach cannot provide answers to the more fundamental ethical questions which this research raises, and be aware that a certain proportion of the population feels strongly that this research is wrong.

Afterword

Discussion of the issues described in this book has been so extensive that one might have thought that attitudes had been set and debate exhausted. However, stimulated perhaps by news about breakthroughs in stem cell research, and anxious to perfect methods of regulation, the House of Commons Science and Technology Committee decided to conduct a review of the relevant law in 2003.

The Department of Health undertook its own investigation in 2004 and, after several exchanges between the interested parties, it was announced that there was to be new legislation reforming the HFE Act and, possibly, merging the HFEA with the Human Tissue Authority in one new large regulatory body. A Joint Committee of both Houses of Parliament was set up to examine the draft Human Tissue and Embryos Bill published in 2007, and major public consultations were launched. Once again, public and political interest was kindled, and the clinicians, ethicists, regulators, and legislators took up their positions.

All the questions we have considered in this book were reopened. There is widespread public debate about whether or not 'standard' IVF (assuming there is such a procedure) should be a matter for a regulator and, if so, which one and how wide the remit of that regulator should be. Sex selection, access to PGD, and the sanctity of the embryo are once again being examined. It is being asked afresh whether a child's need for a father is a valid consideration in deciding whether treatment should be offered to a particular patient; whether it should be permissible to create animal-human hybrid embryos for research; whether donors should be known and that knowledge shared with the extended family that has been created; what to do with frozen embryos when a relationship breaks down and one of the partners withdraws consent to storage and use; whether surrogacy should be more tightly controlled; and whose responsibility is it to try to limit, as far as possible, the number of multiple births resulting from IVF.

All these issues have become live again and there is still no resolution to the problems to which they give rise. One can conclude by realizing that the questions aired in this book are so profound and fascinating, so important to our understanding of who we are and how we conduct ourselves, that answers will never be firm. The debate will be ongoing for years to come. No sooner does it appear that consensus is approaching, than some new development or demand throws fresh light on the issues and calls for further discussion. This book is a contribution to that continuing quest.

Notes

Introduction

1. For example *Evans v Amicus Healthcare* [2005] Fam 1 (a man's right to withhold consent to IVF treatment); *R (on the application of the Assisted Reproduction and Gynaecology Centre) v HFEA* [2002] EWCA Civ 20 (upholding the policy of limiting the numbers of embryos that may be transferred); *R v HFEA ex p Blood* [1999] Fam 151 (permission to use sperm taken without his consent from a dying/dead man). It was held in 2007 that search warrants that the HFEA had obtained in order to search a particular clinic were unlawfully obtained.

Chapter 1

1. Steptoe, PC and Edwards, RG 'Birth after the reimplantation of a human embryo' *Lancet* **2** (1978) 366.
2. Harris, J 'Intimations of Immortality' *Science* **288**(5463) (2000) 59.
3. Wilcox, AJ, Weinberg, CR, et al 'Incidence of early loss of pregnancy' *N Engl J Med* **319** (1988) 189–94.
4. Human Fertilisation and Embryology Authority, *The HFEA guide to infertility* (2006/7).
5. Human Fertilisation and Embryology Authority, *The HFEA guide to infertility* (2006/7).
6. Capanna, E, 'Lazzaro Spallanzani: At the roots of modern biology' *J Exp Zoo* **285**(3) (1999) 178–196.
7. Clarke, GN 'A.R.T. and history, 1678–1978' *Hum Rep* **21** (2006) 1645–50.
8. Henig, RM *Pandora's Baby: How the First Test Tube Babies Sparked the Reproductive Revolution* (Houghton Mifflin, 2004).
9. Leridon, H and Spira, A 'Problems in measuring the effectiveness of infertility therapy' *Fertility and Sterility* **41** (1984) 580–86.
10. Hjelmstedt, A, et al 'Gender differences in psychological reactions to infertility among couples seeking IVF- and ICSI-treatment' *Acta Obst et Gynec Scand* **78**(1) (1999) 42–8.

11. Human Fertilisation and Embryology Authority 'FAQs about treatment. How likely am I to get pregnant after IVF treatment?', available at <http://www.hfea.gov.uk/en/979.html#How_likely_am_I_to_get>.
12. HFEA 'Storage and use of frozen eggs' (HFEA pamphlets, 2003), available at <http://www.hfea.gov.uk/docs/Storage_and_Use_of_Frozen_Eggs.pdf>.
13. Delvigne, A 'Epidemiology and prevention of ovarian hyperstimulation syndrome (OHSS): a review' *Hum Rep Update* **8**(6) (2002) 559–77.
14. Kennedy, R 'Risks and complications of assisted conception' (BFS Factsheets, 2005), available at <http://www.fertility.org.uk/public/factsheets/conceptionrisks.html>.
15. Human Fertilisation and Embryology Authority 'Facts & figures (2006–7)'. (Figures detailing reproductive treatments, risks, success rates etc between 1991 and 2006), available at <http://www.hfea.gov.uk/en/1540.html>.
16. Ford, N, Meister K, et al 'Increased risk of birth defects among children from multiple births' *Birth Defects Research Part A: Clinical and Molecular Teratology* **67**(10) (2003) 879–85.
17. Cox, G F, Burger, J et al 'Intracytoplasmic sperm injection may increase the risk of imprinting defects' *Am J Hum Genet*, **71** (2002) 162–4.
18. *Evans v Amicus Healthcare* [2003] EWHC 2161 (Fam).
19. Pattinson, SD, Caulfield T 'Variations and voids: the regulation of human cloning around the world' *BMC Med Ethics*, **5** (2004) 9, available at <http://www.biomedcentral.com/1472-6939/5/9>.
20. See, for example, Dawkins, R 'What's Wrong with Cloning?' in Nussbaum, MC and Sunstein, CR (eds) *Clones and Clones: Facts and Fantasies about Human Cloning* (New York: WW Norton & Company, 1998).
21. Although doubt has been shed on the feasibility of this since it emerged that the leader in the field, Korean Woo Suk Hwang, had fabricated his results.

Chapter 2

1. The terms 'ethical' and 'moral' are sometimes used interchangeably, although some philosophers and social scientists see them as having specific and distinct meanings. In this book, the former approach has been adopted.
2. Edwards, RG, and Sharpe, DJ 'Social values and research in embryology' *Nature* **231**, (1971) 87–91.
3. Deech, R 'Assisted Reproductive Techniques and the Law' *Med Leg J* **69**(1) (2001) 13, 14.

4. Watson, JD 'Potential consequences of experimentation with human eggs (International scientific and legislative cooperation to protect work on human genetic engineering)' *Comm. on Sci. and Astronaut. Intern. Sci. Policy* (1971) 149–61.
5. Human Genetics Advisory Commission and the Human Fertilisation and Embryology Authority 'Cloning issues in reproduction, science and medicine' (1998), available at <http://www.advisorybodies.doh.gov.uk/hgac/papers/papers_c.htm>.
6. Wheldon, J 'Ethical row over world's first "made to order" embryos' *Daily Mail* (4 August 2006).
7. See the comments for the story cited above, available at <http://www.dailymail.co.uk/pages/live/articles/health/healthmain.html?in_article_id=399142&in_page_id=1774#StartComments>.
8. Human Genetics Commission 'Draft Report on Personal Genetic Information' available at <http://www.hgc.gov.uk/UploadDocs/DocPub/Document/hgc01-p16.pdf>.
9. This approach to embryo research was eventually adopted by Italy, a largely Catholic nation.
10. See, for example, Gillon, R 'Ethics needs principles—four can encompass the rest—and respect for autonomy should be "first among equals"' *J Med Ethics* **29** (2003) 307–12.
11. Mill, JS 'On Liberty' in Warnock, M (ed) *Utilitarianism and On Liberty* (Blackwell, 2003).
12. Chaudhary, V 'Fury at 8 baby birth bonus' *The Guardian* (12 August 1996), available at <http://www.guardian.co.uk/fromthearchive/story/0,,1016421,00.html>.
13. The courts seem to concur with this. See, for example a case of consensual sado-masochism in which the participants were found to have committed a crime: *R v Brown* [1994] 3 AC 212 HL.
14. Ramsey, P 'Shall we reproduce?' *J Am Med Assoc* **220** (1972) 1346, 1480.
15. Spriggs, M 'IVF mixup: white couple have black babies' *J Med Ethics* **29** (2003) 65.
16. De Lacey, S 'Parent identity and "virtual" children: why patients discard rather than donate unused embryos' *Hum Rep* **20** (2005) 1661–9.
17. Human Fertilisation and Embryology Authority, Code of Practice (7th edn, 2007) para 3.1.1.
18. Human Fertilisation and Embryology Authority, Code of Practice (7th edn, 2007) para 3.4.1.
19. Birchard, K 'Challenges to ethics of gamete storage cause headaches in the UK' *Lancet* **355**(9197) (2000) 50.

20. *Mrs U v Centre for Reproductive Medicine* [2002] EWCA Civ 565.
21. Parfit, D *Reasons and Persons* (Oxford: Clarendon Press, 1984).
22. National Bioethics Advisory Commission 'Cloning Human Beings' (June 1997) Chapter 6.
23. See, for example, para 4.7 of the Human Genetics Advisory Commission and HFEA Report 'Cloning issues in reproduction, science and medicine' (December 1998), available at <http://www.advisorybodies.doh.gov.uk/hgac/papers/papers_d.htm>.

Chapter 3

1. 'Science screens out defective genes' *BBC News Online* (18 November 2000).
2. Harmon, A 'Couples Cull Embryos to Halt Heritage of Cancer' *New York Times* (3 September 2006), available at <http://www.nytimes.com/2006/09/03/health/03gene.web.html?ex=1314936000&en=8ac55eefd8dadfd3&ei=5088&partner=rssnyt&emc=rss>.
3. Buxton, J 'Unforeseen uses of PGD—ethical and legal issues' in Horsey, K and Biggs, H (eds) *Human Fertilisation and Embryology: Reproducing Regulation* (Routledge-Cavendish, 2006), 111.
4. Ogilvie, CM, Braude, PR et al 'Preimplantation genetic diagnosis—an overview' *J Histochem Cytochem* **53**(3) (2005) 255–60.
5. King, D 'Preimplantation Genetic Diagnosis and "slippery slopes"' *BioNews* (13 May 2007), available at <http://www.bionews.org.uk/commentary.lasso?storyid=3441>.
6. For a discussion of this, see Marks, D *Disability: controversial debates and psychosocial perspectives* (Routledge, 1999).
7. Spriggs, M 'Lesbian couple create a child who is deaf like them' *J Med Ethics* **28** (2002) 283.
8. Solveig, MR 'Disability, gene therapy and eugenics—a challenge to John Harris' *J Med Ethics* **26** (2000) 89–94.
9. 'We desperately want a girl', *BBC News Online* (12 November 2003) available at <http://news.bbc.co.uk/1/hi/health/3260827.stm>.
10. Human Fertilisation and Embryology Authority 'Outcome of the 1993 consultation—open letter to the Under Secretary of State for Public Health' available at <http://www.hfea.gov.uk/docs/Appendix_B_-_Outcome_of_the_1993_consultation-_open_letter_to_the_Under_Secretary_of_State_for_Public_Health.pdf>.
11. Harris, J 'Sex selection and regulated hatred' *J Med Ethics* **31** (2005) 291–4.

12. Human Fertilisation and Embryology Authority 'Sex selection: report summary', available at <http://www.hfea.gov.uk/docs/Final_sex_selection_summary.pdf>.
13. Siegel-Itzkovich, J 'Israel allows sex selection of embryos for non-medical reasons' *Brit Med J* **330** (2005) 1228.
14. 'Parents seek test tube "lifesaver"' *BBC News Online* (1 October 2001), available at <http://news.bbc.co.uk/1/hi/health/1572106.stm>.
15. *Quintavalle v HFEA* [2005] UKHL 28.
16. This Code of Practice has now been superseded by a later edition in which the relevant section has been substantially cut down. Paragraph 3.1.1 of the 7th edition of the HFEA Code of Practice reads 'The treatment centre should take into account the welfare of any child who may be born as a result of treatment and of any other child who may be affected by the birth before providing any treatment services.'
17. Bhattacharya, S 'Banned "designer baby" is born in UK' *New Scientist* (19 June 2003), available at <http://www.newscientist.com/article.ns?id=dn3854>.
18. 'Quango in designer baby row' *Daily Mail* (18 July 2002), available at <http://www.dailymail.co.uk/pages/live/articles/health/thehealthnews.html?in_article_id=128699&in_page_id=1797>.

Chapter 4

1. Bartlett, K 'Feminism and Family Law' *Fam Law Quart* **33** (1999) 475.
2. Potts, M and Campbell, M 'History of contraception' *Gynaec and Obst* **6** (2002) chapter 8.
3. See Pearsall, R *The Worm in the Bud: the World of Victorian Sexuality* (Weidenfeld & Nicolson, 1969).
4. 'Woman loses final embryo appeal' *BBC News Online* (10 April 2007), available at <http://news.bbc.co.uk/1/hi/health/6530295.stm>.
5. *Davis v Davis* (1992) (Tenn. S. Ct. 1992) Tenn. LEXIS 400.
6. Chen, J 'The Right to Her Embryos: An Analysis of Nahmani v. Nahmani and its Impact on Israeli *In Vitro* Fertilization Law' *Cardozo J Int'l & Comp Law* 7(2) (1999) 325–58.
7. Boseley, S 'Frozen egg baby hailed as fertility milestone' *The Guardian* (11 October 2002), available at <http://www.guardian.co.uk/uk_news/story/0,3604,809837,00.html>.
8. Department of Health 'Research specification—independent evaluation of the 'pharmacy chlamydia screening pathfinder': background & policy context', available at <http://www.dh.gov.uk/en/Policyandguidance/Healthandsocialcaretopics/Chlamydia/index.htm>.

9. Bewley, S, Braude, P, et al 'Which career first?' *Brit Med J* **331** (2005) 588–9, available at <http://www.bmj.com/cgi/content/full/331/7517/588>.
10. 'Many couples unfit for pregnancy' *BBC News Online* (1 September 2005), available at <http://news.bbc.co.uk/1/hi/health/4202630.stm>.
11. Templeton, S-K 'Late motherhood "as big a problem" as teenage mums' *Sunday Times* (13 August 2006), available at <http://www.timesonline.co.uk/tol/news/uk/article607588.ece>.
12. Künzle, R, Mueller, MD 'Semen quality of male smokers and nonsmokers in infertile couples' *Fertility and Sterility* **79**(2) (2003) 287–91.
13. Sharpe, RM 'Environment, lifestyle and male infertility' *Best Practice & Research Clinical Endocrinology & Metabolism* **14**(3) (2000) 489–503.
14. 'Frozen egg birth brings IVF hope' *BBC News Online* (11 October 2002), available at <http://news.bbc.co.uk/1/hi/health/2318609.stm>.
15. 'I had to lie about my age' *The Telegraph* (5 May 2006), available at <http://www.telegraph.co.uk/news/main.jhtml?xml=/news/2006/05/05/nmum05.xml>.
16. 'Rod Stewart confirms new baby due' *BBC News Online* (1 June 2005), available at <http://news.bbc.co.uk/1/hi/entertainment/music/4599259.stm>.
17. Monbiot, G 'Our strange fear of older mothers' *The Guardian* (25 January 2001), available at <http://www.guardian.co.uk/Colomnists/Column/0,5673,427825,00.html>.
18. Bewley, S, Braude, P, et al 'Which career first?' *Brit Med J* **331** (2005) 588–9.
19. Human Fertilisation and Embryology Authority 'How likely am I to get pregnant'?, available at <http://www.hfea.gov.uk/en/979.html>.
20. Scully, JL, Banks, S, et al 'Chance, choice and control: Lay debate on prenatal social sex selection' *Soc Sci & Med* **63** (2006) 21–31.
21. Goodchild, S and Owen, J 'The baby millionaires: Fertility experts become medical profession's highest earners' *The Independent* (8 January 2006).
22. Rapp, R 'Moral pioneers: women, men and fetuses on a frontier of reproductive technology' *Women & Health* **13** (1987) 101–16.
23. 'UK woman killed by rare IVF risk' *BBC News Online* (13 April 2005), available at <http://news.bbc.co.uk/1/hi/health/4440573.stm>.
24. Human Fertilisation and Embryology Authority 'The HFEA Guide to Infertility (2007/8)', available at <http://www.hfea.gov.uk/docs/Guide2.pdf>.
25. Redshaw, N, Hockley, C, et al 'A qualitative study of the experience of treatment for infertility among women who successfully became pregnant' *Hum Rep* **22**(1) (2007) 295–304.
26. Human Fertilisation and Embryology Authority *The HFEA guide to infertility* (2007/8).

27. Franklin, S 'Making Miracles: Scientific Progress and the Facts of Life' in Franklin & Ragoné (eds) *Reproducing Reproduction* (Philadelphia: University of Pennsylvania, 1998).
28. Sherwin, S 'Feminism and Bioethics' in Wolf (ed) *Feminism and Bioethics* (OUP, 1996) 47–66.
29. House of Commons Science and Technology Committee 2005, *Human Reproductive Technologies and the Law*, Report p 49.
30. Nayernia, K, Nolte, J et al '*In Vitro*-Differentiated Embryonic Stem Cells Give Rise to Male Gametes that Can Generate Offspring Mice' *Developmental Cell* 11 (2006) 125–132.
31. Master, Z 'Embryonic stem-cell gametes: the new frontier in human reproduction' *Hum Rep* **21**(4) (2006) 857–63.
32. Human Fertilisation and Embryology Authority 'Horizon scanning briefing: *In vitro* derived gametes' (13 September 2005), available at <http://www.hfea.gov.uk/docs/2006-11-30_SCAG_PAPER_2_-_Prioritisation_of_Horizon_Scanning_Issues_2006.pdf>.
33. McKie, R 'Men redundant? Now we don't need women either' *The Observer* (10 February 2002), available at <http://observer.guardian.co.uk/international/story/0,6903,648024,00.html>.
34. Ikechebelu, JI, Onwusulu, DN, et al 'Term abdominal pregnancy misdiagnosed as abruptio placenta' *Nig J Clin Pract* **8**(1) (2005) 43–5.
35. Pollak, A and Fuiko R [Extremely premature infants—survival and lifespan at the limits of feasibility] *Wiener Klinische Wochenschrift* **117**(9–10) (2005) 305–7.
36. Newson, A 'From foetus to full term—without a mother's touch' *The Times* (30 August 2005), available at <http://www.timesonline.co.uk/tol/news/uk/article560384.ece>.
37. Firestone, S *The Dialectic of Sex: The Case for Feminist Revolution* (New York, 1971) 198–9.
38. 'Hope for spina bifida babies' *BBC News Online* (20 October 1999), available at <http://news.bbc.co.uk/1/hi/health/479416.stm>.
39. Walsh, DS and Adzick, NS 'Foetal surgery for spina bifida' *Seminars in Neonatology* **8**(3) (2003) 197–205.
40. Rothman, BK *The tentative pregnancy: prenatal diagnosis and the future of motherhood* (London: Unwin and Hyman, 1988).
41. Taylor, JS 'Image of Contradiction: Obstetrical Ultrasound in American Culture' in Franklin, S and Ragoné, H (eds) *Reproducing reproduction: kinship, power, and technological innovation* (Philadelphia: University of Pennsylvania Press, 1998).
42. Jackson, E, *Regulating reproduction: law, technology and autonomy* (Oxford: Hart, 2001) 121.

Chapter 5

1. Blood, D *Flesh and Blood, the human story behind the headlines* (Edinburgh: Mainstream Publishing, 2004).
2. Blood, D *Flesh and Blood, the human story behind the headlines* (Edinburgh: Mainstream Publishing, 2004) 52.
3. O'Neill, O 'Informed consent and public health' *Phil Trans R Soc B: Bio Sci* **359**(1447) (2004) 1133–6.
4. Beyleveld, D and Brownsword, R 'Human Dignity, Human Rights, and Human Genetics' *Mod Law Rev* **61**(5) (1998) 661–80.
5. *St. George's Healthcare NHS Trust v S* [1998] 3 WLR 936.
6. See, for example: 'Four charged over US bones theft' *BBC News Online* (23 February 2006), available at <http://news.bbc.co.uk/1/hi/world/americas/4742844.stm>, where the commodification of body parts has been associated with incentives for the unscrupulous to try to profit from selling material from cadavers.
7. Conway, O 'Brain dead US woman has baby girl' *BBC News Online* (3 August 2005), available at <http://news.bbc.co.uk/1/hi/world/americas/4740721.stm>.
8. One of the most bizarre twists of the relationship between the media and the HFEA involved an allegation by the *Daily Mirror* that the HFEA chairwoman was pregnant at the time she was recommending Mrs Blood be denied access to her husband's sperm (Anton Antonowicz 'I'm so lonely' *Daily Mirror* (19 October 1996)). Not only was the suggestion false, but it would have been extremely surprising had it been true, in view of the fact that she was aged 53 at the time.
9. See, for example, 'Widow allowed dead husband's baby' *BBC News Online*, originally published 6 February 1997, available at <http://news.bbc.co.uk/onthisday/hi/dates/stories/february/6/newsid_2536000/2536119.stm>.
10. Riddell, M 'Who in Britain would deny Diane her baby?' *Daily Mirror* (18 October 1996).
11. McLean, S 'Review of the Common Law Provisions relating to the removal of gametes and of the consent provisions in the Human Fertilisation and Embryology Act 1990' (July 1998).

Chapter 6

1. 'Killer serving life "wants baby"' *BBC News Online* (10 January 2007), available at <http://news.bbc.co.uk/1/hi/england/humber/6246975.stm>.

See also: Seamark, M and Dolan, A 'Murderer and his fraudster wife are given £20,000 legal aid to fight for an IVF baby' *Daily Mail* (31 January 2007), available at <http://www.dailymail.co.uk/pages/live/articles/news/news.html?in_article_id=427983&in_page_id=1770>.

2. See, for example, *R v Secretary of State for Home Department, ex parte Mellor* [2001] EWCA Civ 472 (4 April 2001).
3. See Riley, L 'Equality of Access to NHS-funded IVF treatment in England and Wales' in Horsey, K and Biggs, H (eds) *Human Fertilisation and Embryology: Reproducing Regulation* (Routledge-Cavendish, 2007) 83.
4. Harris, J '"Goodbye Dolly?" The Ethics of Human Cloning' *J Med Ethics* 23 (1997) 353–60.
5. Warren, M A 'Sex Selection: Individual Choice or Cultural Coercion?' in Kuhse and Singer (eds) *Bioethics: an Anthology* (Blackwell, 1999).
6. (1998) 39 BMLR 128, [1997] ECHR 20, [1997] Fam Law 605, (1997) 24 EHRR 143, [1997] 3 FCR 341, [1997] 2 FLR 892.
7. Under the Gender Recognition Act 2004 people who have undergone sex changes can now have their new gender recognized in law.
8. United States Supreme Court (1972) *Eisenstadt v Baird* 405 US 438, 453.
9. Chadwick, R (ed) *Ethics, reproduction and genetic control* (London: Routledge, 1987) 4.
10. See *Queen v Louise Collins; Pathfinder Mental Health Services NHS Trust and St George's Healthcare NHS Trust ex parte S* [1998] EWHC Admin 490.
11. Brazier, M 'Reproductive Rights: Feminism or Patriarchy?' in Harris, J and Holm, S (eds) *The Future of Human Reproduction* (Oxford University Press, 1998) 66–76.
12. *Assisted Reproduction and Gynaecology Centre, R (on the application of) v The Human Fertilisation and Embryology Authority* [2002] EWCA Civ 20.
13. Ombelet, W, De Sutter, P, et al 'Multiple gestation and infertility treatment: registration, reflection and reaction—the Belgian project' *Hum Rep Update* 11(1) (2005) 3–14.
14. These figures are taken from Ledger, WL and Anumba, D 'The costs to the NHS of multiple births after IVF treatment in the UK' *BJOG: An International Journal of Obstetrics & Gynaecology* 113(1) (2006) 21–5.
15. Kerr, J and Brown, C, et al 'The experiences of couples who have had infertility treatment in the United Kingdom: results of a survey performed in 1997' *Hum Rep* 14(4) (1999) 934–8.
16. *Supra* note 12.
17. Goodchild, S and Owen, J 'The baby millionaires: Fertility experts become medical profession's highest earners' *The Independent*, 8 January 2006.
18. This perverse incentive has been exacerbated by limitations on the number of state-funded cycles provided. See for example Riley, L 'Access to NHS-funded IVF treatment in England and Wales' in Horsey, K and

Biggs, H (eds) *Human Fertilisation and Embryology: Reproducing Regulation* (Oxford: Routledge-Cavendish, 2007) 99–100.

19. National Bioethics Advisory Commission 'Cloning Human Beings' (June 1997) Chapter 6.

Chapter 7

1. Mieth, D, Haker, H, et al 'European Network for Biomedical Ethics Final Report' (Center for Ethics in the Sciences and Humanities, University of Tübingen, May 1999).
2. Warnock Report, para 2.11.
3. Prime Minister's monthly press conference (27 February 2007), available at <http://www.number10.gov.uk/output/Page11119.asp>.
4. Benson, H *The conflation of marriage and cohabitation in government statistics—a denial of the difference rendered untenable by an analysis of outcomes* (Bristol Community Family Trust, 2006).
5. *Rose & Anor v Secretary of State for Health and Human Fertilisation and Embryology Authority* [2002] EWHC 1593 (Admin).
6. 'Boy tracks his sperm donor father' *BBC News Online* (2 November 2005), available at <http://news.bbc.co.uk/1/hi/health/4400778.stm>.
7. HGC 'Meeting of the genetic services sub-group' (2003).
8. House of Commons Science and Technology Committee, Fifth Report 'Human Reproductive Technologies and the Law' Volume II: Oral and written evidence (Q98) (2005); *cf* Joint Committee on the Human Tissue and Embryos (Draft) Bill, Report Volume II: Evidence (Qs 461–2) (2007).
9. House of Commons Science and Technology Committee, Fifth Report 'Human Reproductive Technologies and the Law' Minutes of Evidence—Volume II (HC 7-II) 30 June 2004 (Q235); Joint Committee on the Human Tissue and Embryos (Draft) Bill, Report Volume II: Evidence (Q705) (2007).
10. See, for example, the story at <http://www.ukdonorlink.org.uk/story3.asp> in comparison with that told at <http://www.ukdonorlink.org.uk/story4.asp>.
11. Golombok, S, and Murray, C 'Social versus biological parenting: family functioning and the socioemotional development of children conceived by egg or sperm donation' *J Child Psychol & Psychiatry* 40(4) (1999) 519–27.
12. 'Embryo with two mothers approved' *BBC News Online* (8 September 2005), available at <http://news.bbc.co.uk/1/hi/health/4225564.stm>.
13. Derbyshire, D 'Scientists seeking to create embryos with three parents' *The Telegraph* (17 October 2004), available at <http://www.telegraph.co.uk/news/main.jhtml?xml=/news/2004/10/18/nembry18.xml>.

14. 'Surrogate mother pushes for adoption' *BBC News Online* (12 August 2001), available at <http://news.bbc.co.uk/1/hi/health/1485494.stm>.
15. O'Boyle, A L, Davis, G D, et al 'Informed consent and birth: protecting the pelvic floor and ourselves' *Am J Obst & Gynec* **187**(4) (2002) 981–3.
16. *'Why Mothers Die 2000–2002' The Sixth Report of Confidential Enquiries into Maternal Deaths in the United Kingdom* (RCOG Press, 2004).
17. Jadva, V, Murray, C, et al 'Surrogacy: the experiences of surrogate mothers' *Hum Rep* **18**(10) (2003) 2196–204.
18. Golombok, S, MacCallum, F, et al 'Surrogacy families: parental functioning, parent-child relationships and children's psychological development at age 2' *J Child Psychol & Psychiatry* **47**(2) (2006) 213–22.
19. Golombok, S, Murray, C, et al 'Non-genetic and non-gestational parenthood: consequences for parent-child relationships and the psychological well-being of mothers, fathers and children at age 3' *Hum Rep* **21**(7) (2006) 1918–24.
20. Serafini, P 'Outcome and follow-up of children born after IVF-surrogacy' *Hum Rep Update* **7**(1) (2001) 23–7.
21. House of Commons Science and Technology Committee, Fifth Report 'Human Reproductive Technologies and the Law' (2005); *cf* Joint Committee on the Human Tissue and Embryos (Draft) Bill Report (2007) 65.
22. See 'IVF "need for father" rule may go' *BBC News Online* (13 July 2006), available at <http://news.bbc.co.uk/1/hi/health/5175640.stm>.
23. In fact, a former head of the HFEA has attacked this clause. See 'IVF "father figure" law attacked' *BBC News Online* (21 January 2004), available at <http://news.bbc.co.uk/1/hi/health/3416055.stm>.
24. This is not, of course, to deny the existence of a 'glass ceiling', and the persisting inequalities between women's and men's incomes. However, such inequalities as exist now are of a different order to the severe constraints which used to exist on women's employment options.
25. Winterman, D 'I'm the daddy' *BBC News Online* (17 May 2006), available at <http://news.bbc.co.uk/1/hi/magazine/4986804.stm>.
26. Copeland, D, and Harbaugh, BL 'Differences in parenting stress between married and single first time mothers at six to eight weeks after birth' *Issues in Comp Ped Nursing* **28**(3) (2005) 139–52.
27. The State of the Nation Report 'Fractured Families' (Social Justice Policy Group, 2006).
28. Aronson, SR and Huston AC 'The mother-infant relationship in single, cohabiting, and married families: a case for marriage?' *J Fam Psychol* **18**(1) (2004) 5–18.
29. Murray, C, and Golombok, S 'Going it alone: solo mothers and their infants conceived by donor insemination' *Am J Orthopsychiatry* **75**(2) (2005) 242–53.

30. Maccallum, F, and Golombok, S 'Children raised in fatherless families from infancy: a follow-up of children of lesbian and single heterosexual mothers at early adolescence' *J Child Psychol & Psychiatry* **45**(8) (2004) 1407–19.
31. Verhulst, FC, Versluis-den Bieman, HO, et al [Being raised by lesbian parents or in a single-parent family is no risk factor for problem behavior, however being raised as an adopted child is], *Ned Tijdschr Geneeskd* **141**(9) (1997) 414–8.
32. See for example, Carlson, M 'Family structure, father involvement, and adolescent behavioural outcomes' *J Marriage and Family* **68**(1) (2006) 137–54.
33. 'One third of births "not planned"' *BBC News Online* (17 November 2006), available at <http://news.bbc.co.uk/1/hi/scotland/edinburgh_and_east/6153736.stm>.
34. House of Commons Science and Technology Committee, Fifth Report 'Human Reproductive Technologies and the Law' (2005).
35. Joint Committee on the Human Tissue and Embryos (Draft) Bill, Report (2007) 65.
36. MacLean, S 'The right to reproduce' in Campbell, T, et al (eds) *Human rights: from rhetoric to reality*. (Oxford: Blackwell, 1986) 99–122. The Equality Act (Sexual Orientation) Regulations 2007 grant exemption to statutory obligations.
37. Many fertility treatments are paid for privately so not all of the arguments related to the conservation of resources apply. However, even privately paid-for treatment may result in pressures on the NHS if, for example, it results in a baby or mother with additional healthcare needs. In these cases, the argument from scarce resources operates at one remove, as it were.
38. See, for example, Carvel, J 'NHS cash crisis bars knee and hip replacements for the obese' *The Guardian* (23 November 2005), available at <http://www.guardian.co.uk/frontpage/story/0,,1648760,00.html>.
39. Galliano, J 'A new kind of family' *The Guardian* (13 August 2003), available at <http://www.guardian.co.uk/gayrights/story/0,12592,1017571,00.html>.
40. Cameron, P 'Children of homosexuals and transsexuals more apt to be homosexual' *J Biosoc Sci* **38**(3) (2006) 413–18; this is supported by Golombok, S, and Tasker, F 'Do parents influence the sexual orientation of their children?' *Developmental Psychology* **32**(1) (1996) 3–11.
41. Anderssen, N, Amlie, C, et al 'Outcomes for children with lesbian or gay parents. A review of studies from 1978 to 2000' *Scand J Psychol* **43**(4) (2002) 335–51.
42. Patterson, C J 'Children of lesbian and gay parents' *Child Dev* **63**(5) (1992) 1025–42.
43. Ahmann, E 'Working with families having parents who are gay or lesbian' *Pediatric Nursing* **25**(5) (1999) 531–5.

44. Lerner, R, Nagai A *No Basis; What the Studies Don't Tell us About Same-Sex Parenting* (Washington DC: Marriage Law Project, 2001).
45. Lee, R G, and Morgan, D, *Human Fertilisation and Embryology* (London: Blackstone Press, 2001) 2.

Chapter 8

1. Mulkay, M *The Embryo Research Debate* (Cambridge University Press, 1997).
2. Mulkay, M *The Embryo Research Debate* (Cambridge University Press, 1997) 26.
3. More recently this approach has been adopted by total opponents of embryo research. Ceasing to rely on moral arguments which have failed to sway the legislators, they claim that there is no evidence that embryo research has succeeded or that it will succeed, and in particular that embryonic stem cell and animal hybrid research will be of no use.
4. Department of Health 'Review of the Human Fertilisation and Embryology Act' Cm 6989 (2006).
5. Bruce, DM 'A Social Contract for Biotechnology: Shared Visions for Risky Technologies?' *J Agri & Env Ethics* **15**(3) (2002) 279–89.
6. The Prime Minister's speech to the Royal Society (2002) *Science Matters*, available at <http://www.number-10.gov.uk/output/Page1715.asp>.
7. Department of Health 'Review of the Human Fertilisation and Embryology Act' (2006).
8. For a discussion of the impact of this tendency, and the problems which arise from discrepancies between scientific policy documents and public concerns, see Parliamentary Office of Science and Technology 'Public dialogue on science & technology' (Report 189, November 2002).
9. Teek, A 'Stem cell therapy is my only chance' *BBC News Online* (17 November 2005), available at <http://news.bbc.co.uk/1/hi/programes/this_world/4438820.stm>.
10. Kirby, A 'Fish do feel pain, scientists say' *BBC News Online* (30 April 2003), available at <http://news.bbc.co.uk/1/hi/sci/tech/2983045.stm>
11. Of course, as embryos develop further, this argument becomes harder to sustain. This is one reason why embryo research is confined to the fourteen-day period which precedes the vestigial development of the brain and spinal cord.
12. Department of Health 'Stem cell research: Medical progress with responsibility' (2000).
13. 'Experts examined—Dr Hwang Woo-suk' *BBC News Online* (8 August 2005) available at <http://news.bbc.co.uk/1/hi/health/4656733.stm>.

14. '"Donor eggs for science" debated' *BBC News Online* (7 September 2006), available at <http://news.bbc.co.uk/1/hi/health/5323894.stm>.
15. The grounds were that the statutory definition did not appear to include embryos created by processes that do not involve the union of egg and sperm. Cell nuclear replacement does not involve sperm in the creation of the embryo: *R v Secretary of State for Health ex p Quintavalle* [2003] UKHL 13.
16. 'Ethical stem cell lines created' *BBC News Online* (24 August 2006), available at <http://news.bbc.co.uk/1/hi/health/5272648.stm>.
17. Pearson, H and Abbott, A 'Stem cells derived from 'dead' human embryo' *Nature* **443** (2006) 376–7.
18. Fangerau, H 'Can artificial parthenogenesis sidestep ethical pitfalls in human therapeutic cloning? An historical perspective' *J Med Ethics* **31**(12) (2005) 733–5.
19. Fang, ZF, Hui, G, et al 'Rabbit embryonic stem cell lines derived from fertilized, parthenogenetic or somatic cell nuclear transfer embryos' *Exp Cell Research* **312**(18) (2006) 3669–82.
20. Denker, HW 'Potentiality of embryonic stem cells: an ethical problem even with alternative stem cell sources' *J Med Ethics* **32**(11) (2006) 665–71.
21. This compromise has been upheld by the House of Lords Select Committee 'Report on Stem Cell Research' (2002) para 4.21.
22. Human Fertilisation and Embryology (Research Purposes) Regulations 2001, SI 2001/188.
23. Embryo Protection Act 1990. See Gottweis, H 'Stem Cell Policies in the US and in Germany' *Pol Stud J* **30**(4) (2002) 444–69.
24. Stem Cell Act, 28 June 2002. See Isasi and Knoppers 'Beyond the permissibility of embryonic and stem cell research: substantive requirements and procedural safeguards' *Hum Rep* **21**(10) (2006) 2474–81.
25. Medically Assisted Procreation Law, Law 40/2004 (19 February 2004).
26. *Planned Parenthood v Casey* 505 US 833 (1992).
27. Public Law Number 144–99, title I, para 128, 110 Stat 26, 34 (1996) governs the situation.
28. Proposition 71, the California Stem Cell Research and Cures Act, was approved by seven million voters (59%) in November 2004.
29. 2004 C.26:2Z-2.
30. Jones, DG, and Towns, CR 'Navigating the Quagmire: or the regulation of human embryonic stem cell research' *Hum Rep* **21**(5) (2006) 1113–16.

Index

A, Mr and Mrs, case 40
Adoption 162–3
Akinbolagbe, Temilola 88, 90
Allwood, Mandy 35–6
Alzheimer's 198
Amir, Yigal 129, 131, 133
Animal research 189
Artificial insemination 15–17
Austria 211
Australia 212
Autonomy 34–43, 91–4

Barlow, Tony 175
Beasley, Helen 159–60
Belgium 212
Best interests 104, 109–17
Blair, Tony 154, 181
Blood, Diane and Stephen 101–26, 217
Brazier, Margaret 138
Brown, Louise 7, 9, 181
Bush, President 210
Buttle, Liz 84–7

California 210–11
Canada 212
Cancer 56
Cloning 10, 25–7, 49, 202
  therapeutic 28, 185, 200
Code of Practice, HFEA
  PGD 54–5, 74
  welfare of the child 43, 67, 169–71, 177, 221
Cohabitation 155
Consent 103, 106
Council on Bioethics 210
Cystic Fibrosis 54, 61

Davis, Junior 79, 221
Dead and comatose patients 115–21
Diabetes 55
Dickson, Kirk 127–33
Dignity 108
Disability 57–60
Discrimination 172–6
Divorce 155
DNA 26–8, 158
Dolly the sheep 10, 25–7
Donaldson, Mary 8
Donor anonymity 157, 165
Donor parents 156
Down syndrome 45–6
Drewitt, Barrie 175
Duchenne muscular dystrophy 46
Duchesneau, Sharon 59–60

Edwards, Robert 8, 29, 40, 217–18
Egg 23, 123, 197–8
Eggs, frozen 24, 44–5, 81
Electro-ejaculation 107, 109
Embryos
  definition 201–3
  development 18–20
  ethical concerns 188–96
  research on 179–82, 198–9
  sale of 31
  surplus 41–2, 196
Ethics 4, 29–38, 50–1
European law 120, 125, 149–52, 208
Evans, Natallie 78–80, 217–18

Family 135–40, 153–5, 176–8
Father, need for 165–71
Feminism 77–87
Finland 212

Gametes
 artificial 95, 153
 donated 155–64, 196
Germany 34, 206–8, 211
GM food 180–1
Greece 212

H, Mrs, case 142, 145, 147, 149
Harris, John 134, 217, 220
Hashmi family 65–74
HFE Act 1990 1, 91, 147
 consent 40–1
 counselling 39
 embryo research 188–9, 202
 import of gametes 119–20
 passage 9
 PGD 74
 reform 75, 215
 welfare of the child 43, 154, 165, 170, 177
HFEA 7, 88
 cloning 28, 188
 consent 103, 122–3
 egg freezing 45
 embryo research 195, 201–6
 embryos used 143–8
 future developments 11
 gametes, export of 119–20, 125–6, 198
 PGD 67, 70–5
 public opinion, and 4, 6, 30–2, 50–1, 74, 123–4
 remit 1, 16–17, 42, 101, 150
 sex selection 62–4
HLA 5, 23–4, 43, 68, 71
Human Genetics Commission 33, 219
Human Genetics Advisory Commission 28, 30, 219–20
Human Reproductive Cloning Act 10
Human rights 2–3, 10, 32, 67–8, 91, 104, 127–50
Hungary 212
Huntington's disease 45, 54–5, 61
Hwang, Woo Suk 197, 218

ICSI 9, 21, 42
Infertility 14–15, 173–4
Ireland 211
Israel 212, 221
Italy 209, 219
IVF
 consent 40
 cost 96, 148
 public opinion 10
 restrictions 207, 209, 212
 regulation 215
 risks 20–1, 42, 45, 90
 success rate 39, 89
 technique 8–9, 15, 17–20, 23

Japan 212
Jones, DG 211

Kant, Immanuel 34–5
Kass, Leon 210
Kingsbury, Chad 56

Lockwood, Gillian 83

Marry, right to 140
Masterton family 62–4
McCullough, Candy 59–60
McLean, Sheila 126, 172
Mellor, Gavin 133
Mill, John Stuart 35, 37
Mitochondria 158–60
Mucklejohn, Ian 166–70
Mulkay, Michael 179–80
Multiple births 144–7, 215

Nahmani, case 79–80
National Health Service 3, 39, 96, 142, 144, 146, 148
Neill, Carolyn 44–5
Netherlands, the 212
New Jersey 210–11
Norway 211

OHSS 21, 88
Older mothers 81–7

Pancoast, William 16
Parkinson's 198–9, 203
Parthenogenesis 204–5

Perry, Helen 81
PGD 22–3, 42–3, 53–62, 65–75
PGS 22
Poland 211
Precautionary principle 150–2, 164
Pregnancy 97–9
Pro-Life Alliance 202

Rashbrook, Patricia 84–7
Regulation 206–13
Reproduction 11–22
Rights, human 2–3, 10, 32, 67–8, 91, 104, 127–50

Same sex parents 175–6
Saviour siblings 65–75
Sex selection 62–5, 215
Sickle cell anaemia 61
Singapore 212
Singer, Peter 193
Single parents 165–70
South Korea 212
Stem cells 10, 28, 95–6, 182–8, 196, 199–204, 206–8, 210, 212–13
  bank 202
Steptoe, Patrick 8, 40, 217
Surrogacy 98, 159–64
Sweden 212

Taiwan 212
Taranissi, Mohamed 142–3, 147
Towns, CR 211
Tricarico, Stefano 186
Triplets 21, 168
Twins 21, 25, 40, 144

U, Mr and Mrs, case 48
USA 209–11

VLA 8

Warnock Report 8–9, 33, 154, 177, 179, 181–2, 206
Warren, Mary Ann 135
Watson, James 30, 219
Welfare of the child 44, 47, 170–1
Whitaker family 71–4
Womb, artificial 97–100

X, Y, Z, case 136